List of Contributors

K. G. M. M. Alberti (*University Department of Chemical Pathology, Southampton General Hospital, Southampton SO9 4XY, U.K.*)

D. G. Beevers (*M.R.C. Blood Pressure Unit, Western Infirmary, Glasgow* G11 6*NT, U.K.*)

J. J. Brown (*M.R.C. Blood Pressure Unit, Western Infirmary, Glasgow* G11 6*NT, U.K.*)

R. Fraser (*M.R.C. Blood Pressure Unit, Western Infirmary, Glasgow* G11 6*NT, U.K.*)

J. R. Hobbs (*Department of Chemical Pathology, Westminster Medical School, London SW*1*P* 2*AR, U.K.*)

H. A. Krebs (*Metabolic Research Laboratory, Nuffield Department of Clinical Medicine, Radcliffe Infirmary, Oxford OX*2 6*HE, U.K.*)

D. Kremer (*M.R.C. Blood Pressure Unit, Western Infirmary, Glasgow* G11 6*NT, U.K.*)

A. F. Lever (*M.R.C. Blood Pressure Unit, Western Infirmary, Glasgow* G11 6*NT, U.K.*)

J. J. Morton (*M.R.C. Blood Pressure Unit, Western Infirmary, Glasgow* G11 6*NT, U.K.*)

I. W. Percy-Robb (*University Department of Clinical Chemistry, The Royal Infirmary, Edinburgh EH*3 9*YW, U.K.*)

J. I. S. Robertson (*M.R.C. Blood Pressure Unit, Western Infirmary, Glasgow* G11 6*NT, U.K.*)

M. A. D. Schalekamp (*M.R.C. Blood Pressure Unit, Western Infirmary, Glasgow* G11 6*NT, U.K.*)

P. F. Semple (*M.R.C. Blood Pressure Unit, Western Infirmary, Glasgow* G11 6*NT, U.K.*)

A. Wilson (*M.R.C. Blood Pressure Unit, Western Infirmary, Glasgow* G11 6*NT, U.K.*)

H. F. Woods (*M.R.C. Clinical Pharmacology Unit and University Department of Clinical Pharmacology, Radcliffe Infirmary, Oxford OX*2 6*HE, U.K.*)

Preface

The last three decades have witnessed a rapid growth in the applications of biochemistry to medicine. Not only have the advances in the knowledge of metabolic pathways and molecular biology increased understanding and hence prevention and treatment of disease, but the use of biochemical techniques for the benefit of individual patients has increased beyond all expectation. The logarithmic rate of growth of medical biochemistry continues. This growth, together with the breadth and depth of the subject, creates difficulties for those who wish to acquaint themselves with new knowledge outside their own specialist area as well as for those seeking to enter the field of medical biochemistry as students.

The Biochemical Society and The Association of Clinical Biochemists thought it opportune, therefore, to launch a new series of volumes entitled *Essays in Medical Biochemistry*, which are intended to be complimentary to the well-established series *Essays in Biochemistry*. The purpose of the new series is to cover those aspects of biochemical knowledge that are of particular relevance to medicine. The Editors' intentions are to commission essays by acknowledged world experts on those biochemical subjects that have undergone rapid growth and expansion during the past few years. In style the emphasis has been placed upon the production of an essay that is pleasant to read and relevant rather than upon one that is an exhaustive review of the subject matter. As far as possible each volume of *Essays in Medical Biochemistry* will contain four or five essays, and it is hoped to include from time to time essays dealing with the applications of new analytical procedures. The object of the latter will be to draw attention to new investigative possibilities as they emerge.

Although the applications of biochemistry to medicine are most obvious in the departments usually known as clinical biochemistry, clinical chemistry or chemical pathology, most branches of medicine have been influenced by the subject. It is not, therefore, the Editors' intention to restrict the subject matter of this series to topics traditionally concerning the routine hospital biochemical service. Every opportunity will be taken to complement *Essays in Biochemistry* by emphasizing the intercellular integrative aspects of biochemistry as they relate to man. Nevertheless, in their choice of subjects the Editors have in mind particularly the needs of hospital biochemists and clinicians with an interest in metabolic diseases. They hope that the essays will be of considerable interest to senior medical and biochemistry students as well as their teachers. It is a matter of great satisfaction to the Editors that they were able to include in their first series of essays by internationally famous authors one under the co-authorship of Professor Sir Hans Krebs. Sir Hans's career, which has been described in detail elsewhere [H. L. Kornberg (1968) H. A. Krebs: a pathway in metabolism. *Biochem. Soc. Symp.* **27**, 3–9] illustrates so well the advances in medical science that can occur when the problems posed by disease are translated into action in the laboratory. We wish to dedicate this first volume to Sir Hans, not only as a token of our own affection and esteem, but also in the hope that his career, and in a lesser way this series, will inspire others to be motivated in the same direction.

V. MARKS
Department of Biochemistry,
University of Surrey,
Guildford,
Surrey GU2 5XH,
U.K.

C. N. HALES,
Department of Medical Biochemistry,
Welsh National School of Medicine,
Heath Park,
Cardiff CF4 4XN,
U.K.

Contents

Essays Med. Biochem. **1**, 1–58
Printed in Great Britain

Recent Developments in the Biochemical Investigation of Hypertension

By D. G. BEEVERS, J. J. BROWN, R. FRASER, D. KREMER, A. F. LEVER, J. J. MORTON, J. I. S. ROBERTSON, M. A. D. SCHALEKAMP, P. F. SEMPLE and A. WILSON

M.R.C. Blood Pressure Unit, Western Infirmary, Glasgow G11 6NT, U.K.

Introduction

Primary abnormalities of the nervous system, heart, blood vessels, sodium, water, corticosteroids, the renin–angiotensin system, catecholamines and the psyche have all at one time or another been considered important in the pathogenesis of hypertension. Nevertheless, despite extensive study the mechanism of hypertension is understood in three rare syndromes only: phaeochromocytoma, primary hyperaldosteronism and renin-secreting tumour. Cushing's syndrome may be a further example. In each of these excess of a pressor substance is released from a neoplastic or hyperplastic gland. The search for excess of other pressor agents in the commoner hypertensive syndromes has been largely fruitless. Possibly such abnormalities do not exist and an explanation must be sought elsewhere.

Mechanisms controlling blood pressure are interrelated; normally, overactivity of one is followed by compensatory underactivity of others, and blood pressure does not rise. As will be discussed, failure of these compensatory adjustments could raise blood pressure to a pathological level without any pressor mechanism deviating from its normal range.

The present review is concerned with recent developments in the biochemical investigation of hypertension; it is in two main parts: the first deals with the contribution of the biochemists to diagnosis, the second with the role of biochemical abnormalities in aetiology. No attempt has been made to cover the extensive literature on earlier work or on recent studies of the morbid anatomy, epidemiology, drug therapy and management of hypertension. Information on these can be obtained from the general texts of Pickering (1, 2) and Kaplan (3) and from recent symposia (4–8) and reviews (9–12). References cited in the present text are mainly reviews and recent papers.

PART I

CLASSIFICATION AND BIOCHEMICAL ASPECTS OF DIAGNOSIS

1. Definitions: Hypertension as a State of Risk

Hypertension is not easily defined; no one disputes that high blood pressure is dangerous and treatable and that the risk decreases when blood pressure is controlled (1, 13, 14). The problem arises when an attempt is made to distinguish normal and abnormal pressure. Hypertension is undoubtedly common. If it were also a disease entity, a frequency distribution curve for blood pressure in a large population should show two subpopulations, one with and the other without high blood pressure. In fact, most surveys show a single uninterrupted curve; Fig. 1 illustrates fairly typical data from a recent Scottish

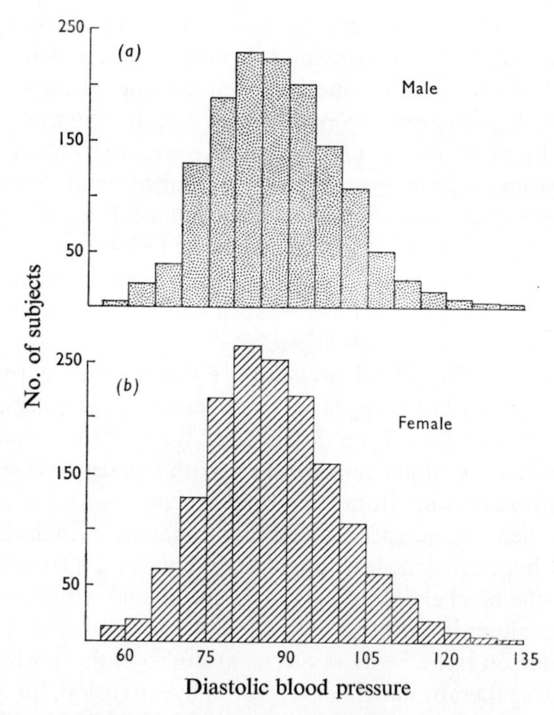

Fig. 1. *Distribution of diastolic blood pressure in 'normal' men and women aged* 45–65 *in the Renfrew screening survey* [*Hawthorne et al.* (15)]

(Reproduced by permission of the *British Medical Journal.*)

Table 1. *'Prevalence' of hypertension by using different diagnostic criteria*

The number of subjects screened was 3061. Values for men (1440 in number) and women were similar. From Hawthorne *et al.* (15).

Diastolic pressure		
90mmHg or more	95mmHg or more	100mmHg or more
39.7%	26.0%	15.6%

study (15). Epidemiological and insurance company data show a graded relation between blood pressure and the risk of subsequently developing cerebrovascular, coronary and renal disease; the relation extends well into the 'normal' range of blood pressure (1, 13).

Observations of this sort support quite strongly Pickering's (1) contention that hypertension is not a qualitative entity in the sense that it is present or absent. It follows that a dividing line drawn between normal and abnormal blood pressure is arbitrary, having no epidemiological basis (Fig. 1). Nor is there a pathological or biochemical basis for the distinction. Indeed, biochemical investigation of hypertensive patients may be compromised by a large overlap between data from 'normal' and 'hypertensive' individuals and this may simply reflect the fact that the two groups are drawn from different parts of the same population.

Nevertheless, a dividing line is needed in the management of hypertension. It is important to identify a level of pressure above which treatment is beneficial. There is a good case for treatment of patients with diastolic pressures of 110mmHg or more (14) and there are plans to determine whether treatment is justified with pressures as low as 90mmHg (14, 16). If it is, more than 20% of the adult population may be eligible for treatment (Table 1). This is only a very approximate value as minor changes of diagnostic criteria produce major changes in the yield of hypertensive patients (Table 1). Other factors to be borne in mind are the marked variability of blood pressure (1) and the liability of the observer to bias in measuring blood pressure (1, 10).

In summary, arbitrary criteria have to be used in distinguishing normal and abnormally high blood pressure. Confusion over the distinction has arisen because a genuine need for treatment has been wrongly taken by some to imply the existence of a distinct pathological entity.

2. Classification of Hypertension

A commonly used classification of hypertensive patients is shown in Table 2. Pickering (1) and Kaplan (3) give more comprehensive lists.

No cause is found for increased pressure after investigation in 70–90% of patients who are consequently considered to have 'essential' hypertension.

Table 2. *Classification of hypertension*

(*a*) Essential hypertension

(*b*) Secondary hypertension
- (i) Renal: Renal vascular disease
  - Renin-secreting tumour
  - Glomerulonephritis
  - Pyelonephritis
  - Polycystic kidney
  - Collagen disease
- (ii) Adrenal cortex: Cushing's syndrome
  - Primary hyperaldosteronism
  - Deoxycorticosterone excess
  - Congenital adrenal hyperplasia
- (iii) Adrenal medulla: Phaeochromocytoma
- (iv) Coarctation of the Aorta
- (v) Hypertensive disease of pregnancy

The term does not imply the existence of a single cause and because the diagnosis is made by a process of exclusion, essential hypertension is considered in more detail in the final section of Part I.

Secondary hypertension can be subdivided into a number of syndromes. Estimates for the prevalence of these vary markedly from one centre to another. Selection is probably important here since patients with a particular syndrome tend to gather in clinics interested in the condition. This is shown in extreme form in the data of Table 3. Primary hyperaldosteronism was found to be common in a blood pressure clinic specializing in the condition and exceedingly uncommon in an unselected series of patients detected in a screening survey by the same clinicians using the same methods. Most estimates of the prevalence of primary hyperaldosteronism are based on hospital clinic experience and different degrees of selection could well account for the varied experience so far published (see ref. 3).

3. Catecholamines

The three main catecholamines, adrenaline, noradrenaline and dopamine are synthesized in nervous tissue; dopamine and noradrenaline are both found

Table 3.

	M.R.C. Blood Pressure Unit (Glasgow)	Renfrew Screening Survey
Total number of hypertensives seen in 1972	134	252
Number confirmed as suffering from primary hyperaldosteronism	18	0
Percentage of hypertensive patients suffering from primary hyperaldosteronism	13	0

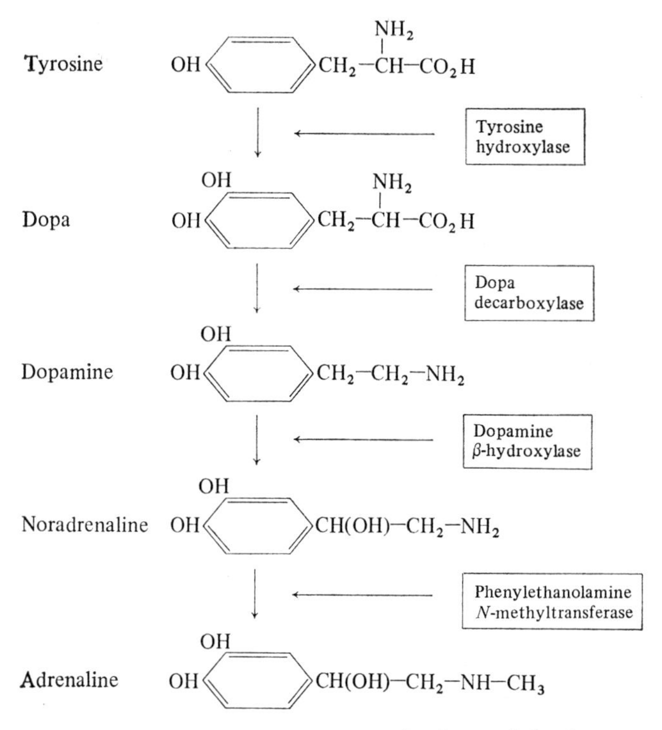

Fig. 2. *Main biosynthetic pathways for the catecholamines*

in the central nervous system but only noradrenaline in the autonomic nervous system (17, 18). Adrenaline, the main product of the adrenal medulla is derived ultimately from tyrosine by two hydroxylation steps, a decarboxylation and a final methylation (Fig. 2). Other synthetic pathways are discussed by Molinoff & Axelrod (17). Corticosteroids seem to be important in the final methylation step (19), this is of some teleological interest since it might account for the close anatomical relation of the adrenal medulla and cortex and for the unusual portal system which carries venous blood from the cortex through the medulla on its way to the heart (20).

The three amines are either excreted as such in urine or degraded by monoamine oxidase and catechol *O*-methyltransferase before excretion (Fig. 3). The relative proportion of the free amines, metadrenalines and vanillylmandelic acid in urine is approximately 1:10:100.

Noradrenaline is a stimulant of α-adrenergic receptors (21) and infusion of the amine in man results in vasoconstriction with pallor, decrease in muscle blood flow and risc of blood pressure. The diastolic pressure increases markedly and the pulse rate falls reflexly because of increased arterial pressure (22, 23). Infusion of adrenaline, with its mixed α- and β-adrenergic effects, produces

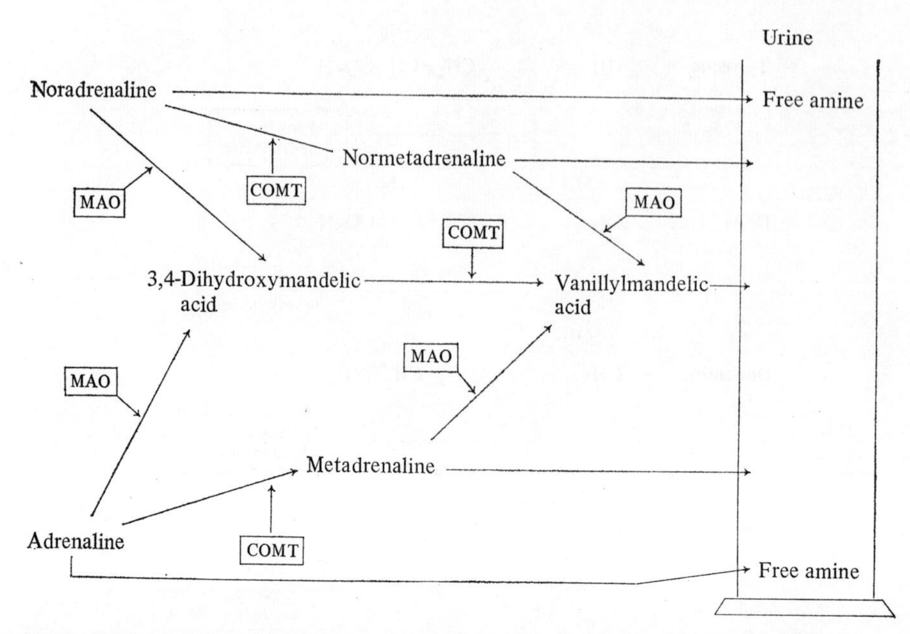

Fig. 3. *Breakdown of catecholamines by monoamine oxidase (MAO) and catechol O-methyltransferase (COMT) and subsequent urinary excretion*

tachycardia, increased cardiac output, headache, anxiety, sweating and glycosuria. Systolic pressure rises but diastolic pressure may fall (22). Palpitation and arrhythmia sometimes develop because of increased myocardial excitability. Dopamine has a mild pressor effect on infusion. Pulse rate and cardiac output rise, but peripheral resistance decreases (22).

4. Phaeochromocytoma

Pathological, clinical and therapeutic aspects of the condition are well covered in recent publications (23, 24, 25). Phaeochromocytomata are tumours of sympathetic nervous tissue; about 90% arise in the adrenal medulla and most of the remainder are to be found within the abdomen (24, 25). The main products of medullary and extramedullary tumours are adrenaline and noradrenaline respectively (25). Malignant phaeochromocytomata are rare (26, 27); some produce mainly dopamine (27). Separate estimation of adrenaline, noradrenaline and dopamine is of value therefore in diagnosis and location of the tumour (25).

(*i*) *Clinical features*

Symptoms and signs of phaeochromocytoma are due to excess catecholamines and can be reproduced by infusion of adrenaline and/or noradrenaline.

Characteristically they include attacks of hypertension with headache, palpitation, tachycardia (when adrenaline predominates), sweating, nausea, anxiety and glycosuria (25). Hypertension is followed in some patients by a period of hypotension but the elevation of blood pressure can be persistent (25).

Familial phaeochromocytomata occur, sometimes in association with medullary carcinoma of the thyroid (28). Another association is the neurocutaneous syndrome, with multiple skin and mucosal neuromata (29).

Symptoms highly suggestive of phaeochromocytoma can develop when patients receiving monoamine oxidase inhibitors eat foods such as Marmite and cheese which are rich in tyramine (30). Similar symptoms occur when clonidine treatment is withdrawn in a hypertensive patient, because of catecholamine release (31).

The first problem for the clinician is to consider the possibility of phaeochromotycoma. Diagnostic delays are common, possibly because the syndrome is rare; its varied clinical presentation and superficial resemblance to an anxiety state are sources of confusion (25).

(*ii*) *Biochemical diagnosis*

Diagnosis is less difficult for the biochemist since distinctly increased urinary excretion of free catecholamines, metadrenaline and vanillylmandelic acid (Fig. 3) can usually be demonstrated either in 24h collections of urine or in shorter collections taken across an attack (25). Plasma levels of catecholamines are also increased (32, 33). Bioassay (25), fluorimetry (34, 35), isotope derivative (32, 33, 36) and g.l.c. (37) techniques are used to measure adrenaline and noradrenaline.

Diagnosis can be taken further, in the difficult case, by precipitating an attack with histamine, tyramine or glucagon, but provocative tests of this type are not now favoured because of their danger and inaccuracy (23). Successful control of blood pressure, by using a combination of α- and β-adrenergic blocking drugs, is also a diagnostic clue (23). As indicated above, a high concentration ratio of adrenaline to noradrenaline suggests a tumour in or close to the adrenal gland. Where the tumour cannot be located by the radiologist, samples of blood can be taken at different levels of the inferior vena cava to determine the entry point for venous blood with a high amine content (25).

False positive catecholamine results may follow excessive intake of certain fruit, particularly bananas. Eating vanilla products sometimes increases vanillylmandelic acid alone. Monoamine oxidase inhibitors, by blocking one of the two pathways of degradation for the amines (Fig. 3), raise metadrenaline and depress vanillylmandelic acid. False positive metadrenaline and vanillylmandelic acid results may arise from treatment with methyldopa. False negative results are produced by reserpine which greatly depresses the synthesis and storage of catecholamines (25).

(iii) Hypertension with phaeochromocytoma

Although the rise of blood pressure during a typical paroxysm is associated with a distinct increase of plasma catecholamines (38), hypertension is sometimes found with normal catecholamine values. Also, successful removal of a phaeochromocytoma occasionally fails to decrease blood pressure. As is discussed below, it is not uncommon in man and animals for hypertension to persist after successful removal of its initial cause.

(iv) Summary

Phaeochromocytoma is a rare, but clear, example of hypertension resulting from excessive release of pressor substances. Adrenaline and noradrenaline are usually responsible and, once the clinical diagnosis has been considered, biochemical confirmation usually presents no problem, though location of the tumour can be difficult.

5. Corticosteroids

(i) Biosynthetic pathways

The adrenal cortex synthesizes and secretes a variety of biologically active steroids affecting both electrolyte and carbohydrate metabolism. Each corticosteroid hormone possesses both of these properties, although one or other usually predominates and the hormone may then be classified as mineralocorticoid or glucocorticoid. In man, aldosterone, corticosterone and deoxycorticosterone behave mainly as mineralocorticoids, whereas cortisol and 11-deoxycortisol act mainly as glucorticoids. Though less certain, 18-hydroxydeoxycorticosterone (39) and possibly 18-hydroxycorticosterone are weak mineralocorticoids.

Biosynthesis follows two main pathways (Fig. 4): the zona fasciculata produces mostly cortisol, 11-deoxycortisol (substance S), corticosterone (substance B), deoxycorticosterone and 18-hydroxydeoxycorticosterone, whereas the zona glomerulosa produces mostly corticosterone, deoxycorticosterone, 18-hydroxycorticosterone and aldosterone. The two zones respond to different stimuli and it is important to note the dual source of corticosterone and deoxycorticosterone.

Cholesterol is a common precursor for corticosteroids. It is converted to progesterone (Fig. 4) by removing most of the side chain at A, by oxidation at B and by isomerisation at C. These are rate-limiting steps, susceptible to control by ACTH* and other stimuli. Subsequently, progesterone undergoes sequential hydroxylation, catalysed by mixed function oxidases (or 'hydroxylases') requiring cytochrome *P*-450 as a cofactor and with a high degree of specificity for the position and orientation of the hydroxyl group within the molecule (40).

In the zona fasciculata, most of the progesterone is converted into the 17-

* Abbreviation: ACTH, adrenocorticotrophin.

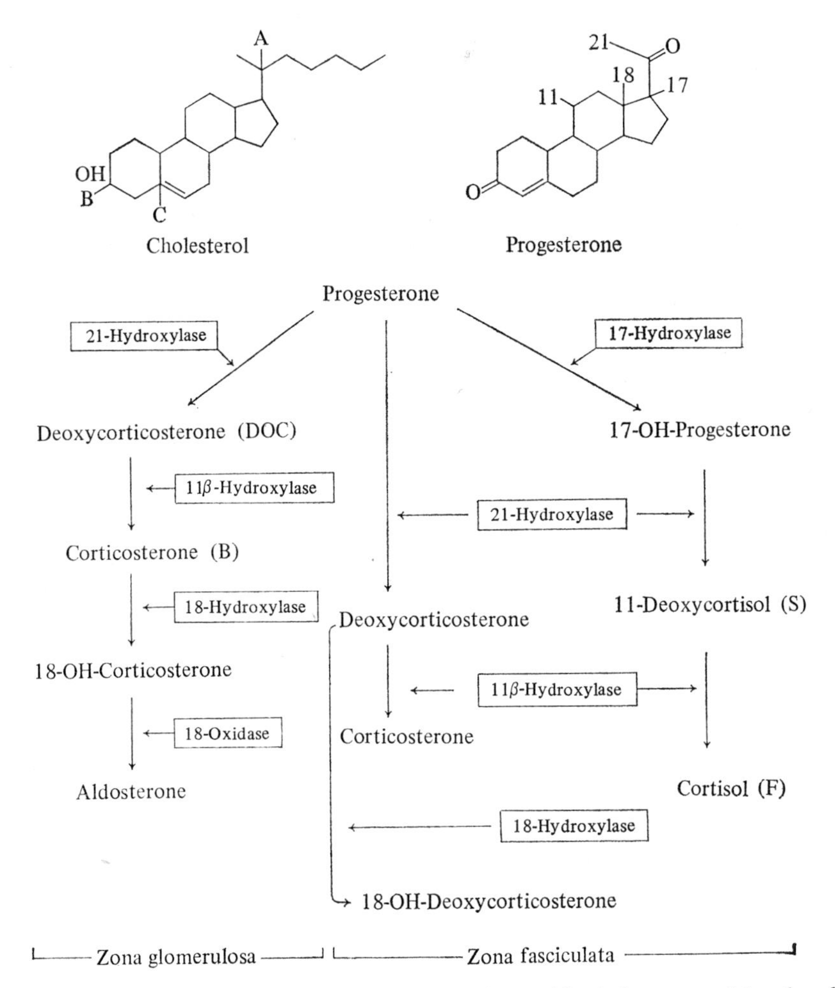

Fig. 4. *Main biosynthetic pathways within the glomerulosa and fasciculata zones of the adrenal cortex*

hydroxy derivative, subsequent 21- and 11-hydroxylations of which yield cortisol (Fig. 4). Equivalent 21- and 11-hydroxylation, without prior 17-hydroxylation, converts progesterone to deoxycorticosterone and corticosterone.

Cells of the human zona glomerulosa contain no 17-hydroxylase activity (24) and therefore convert progesterone to deoxycorticosterone and corticosterone directly (Fig. 4). Oxidation of the 18-methyl group of progesterone is accomplished by two enzyme systems. One is confined to the zona fasciculata and converts deoxycorticosterone to 18-hydroxydeoxycorticosterone (Fig. 4); the other, present only in the zona glomerulosa, forms 18-hydroxycorticosterone which is then specifically oxidized to the unique 18-aldehyde group of aldosterone (41). Other biosynthetic pathways may exist and evidence for these is reviewed by Vecsei & Glaz (42).

(*ii*) *Stimulation of corticosteroid secretion by ACTH*

Hypophysectomy, or suppression of ACTH secretion by dexamethasone, leads to a fall in the secretion of cortisol, corticosterone, deoxycorticosterone and 18-hydroxydeoxycorticosterone. Aldosterone secretion is little affected. Conversely, injection of ACTH results in a rapid increase in the secretion of cortisol and deoxycorticosterone (Fig. 5). As might be expected from the smaller number of hydroxylation steps necessary, plasma concentrations of deoxycorticosterone increase more rapidly than cortisol (Fig. 5). Aldosterone secretion may increase with large doses of ACTH but, at physiological concentrations, the effect is short-lived even during continuous infusion. Stimuli to aldosterone are dealt with in more detail below.

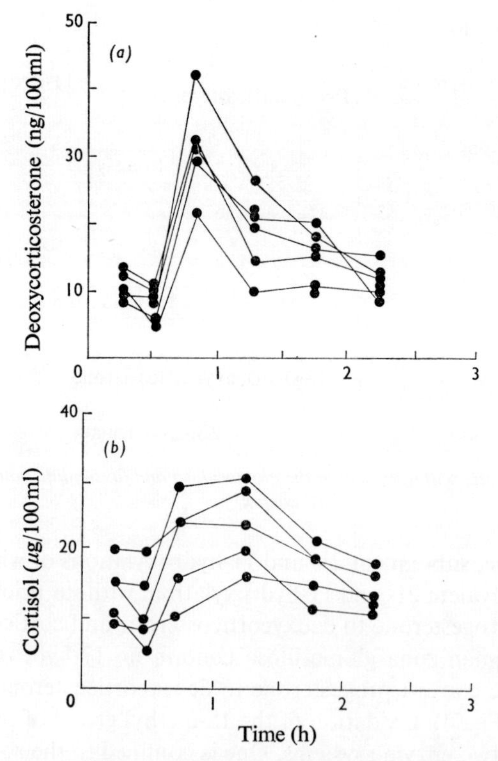

Fig. 5. *Changes in the plasma concentrations of deoxycorticosterone and cortisol in five normal subjects following intravenous injection of tetracosactrin* (1 μg) *immediately after the second plasma sample*

ACTH secretion is stimulated by stress of various types including insulin-induced hypoglycaemia (43). The effect of stress is mediated by corticotrophin releasing factor, a polypeptide of unknown composition, which passes from the hypothalamus through the hypophyseal portal system to stimulate release of ACTH from the anterior pituitary (43, 44). The mechanism is inhibited by high plasma concentrations of cortisol (infused or endogenously produced) and dexamethasone. The 11-deoxycorticosteroids do not operate the feedback mechanism and thus inhibition of the 11-hydroxylation enzymes by metyrapone (metapyrone) decreases cortisol production, stimulates ACTH release and raises the concentration of 11-deoxycortisol and deoxycorticosterone.

The stimulant effect of ACTH on adrenal steroid production is probably exerted on the rate-limiting conversion of cholesterol into progesterone (45, 46) and its effects are thought to be mediated by the adenylate cyclase system (47–49).

(iii) Adrenocorticotrophic hormone

ACTH is a polypeptide with 39 amino acids, the *N*-terminal sequence of 18 (α1–18) residues giving full biological activity. A synthetic preparation of the α1–24 form of ACTH (tetracosactrin) is in common use as a diagnostic and therapeutic agent. The remainder of the polypeptide chain, the *C*-terminal, varies from species to species and its presence may delay catabolism and thereby prolong the biological effect of the *N*-terminal sequence (50).

Until the development of radioimmunoassay (reviewed in ref. 51), measurement of ACTH by bioassay depended upon its stimulating effect on corticosteroid production (43). Most antibodies currently used in the radioimmunoassay seem to recognize the *C*-terminal sequence in the molecule and there may be some discrepancy between assayed and biologically active hormone (52). Radioimmunoassay is sufficiently sensitive to measure at least the upper part of the normal range of ACTH in plasma (52), although more recently, in an attempt to improve specificity and sensitivity, there has been a return to bioassay using the long-recognized ability of ACTH to affect the redox potential of the adrenal gland (53, 54).

6. Cushing's Syndrome

Although hypertension is relatively common in Cushing's syndrome (55), patients usually present to endocrinologists because of the associated diabetes mellitus, obesity and amenorrhoea. Weakness, muscle wasting, bone pain, osteoporosis, and psychiatric disorders may take them to other specialists (52, 55, 56, 57).

(i) Pathology

At least four forms of the syndrome are recognized, each characterized by excess of cortisol or synthetic glucocorticoid.

(*a*) *Pituitary-dependent bilateral adrenocortical hyperplasia*. This is the largest group. The basic abnormality is thought to be excessive production of ACTH, possibly owing to an underlying hypothalamic disorder (58). Circulating levels of ACTH are increased and its circadian rhythm (normally highest on waking and lowest in the evening) is abnormal (52, 59). In a small number of patients the ACTH derives from a radiologically apparent tumour of the pituitary. These have Cushing's disease as originally described. More often, there is nothing to suggest a tumour at first although small tumours may be found later at operation or autopsy (56, 60).

(*b*) *Ectopic ACTH syndrome*. Cortisol excess may also result from secretion of an ACTH-like peptide by a tumour of non-endocrine tissue such as oat-cell bronchial carcinoma (52, 57, 59). This is one of the increasing number of ectopic hormone syndromes to be recognized. The subject has been reviewed recently (61, 62) and is of great interest to clinicians and biochemists.

(*c*) *Tumours of the adrenal cortex*. Cortisol excess can result from tumours of the adrenal cortex. Most of these are benign, the malignant variety being most common in children. Symington (24) has given a full account of the pathological aspects.

(*d*) *Iatrogenic Cushing's syndrome*. Florid Cushing's syndrome can result from long-term treatment with corticosteroids or ACTH. Dermatological preparations containing steroids can produce systemic effects, but prolonged depression of endogenous pituitary–adrenal function seems unusual (63).

(ii) Clinical features

Clinical aspects of diagnosis and management have been fully reviewed (52, 55, 57, 64). The distinction of Cushing's syndrome from simple obesity is sometimes difficult (64, 65) but, in a florid case, obesity of the trunk with muscle wasting in the limbs, striae, hirsuitism, telangectasia, bruising, thin skin, hypertension and glycosuria combine to make the diagnosis of Cushing's syndrome obvious. Hypertension sometimes occurs with only minimal physical signs and in the ectopic ACTH syndrome signs of Cushing's syndrome may be absent, despite gross excess of cortisol, possibly because of the shorter clinical course (52). The prognosis for untreated Cushing's syndrome is poor (56).

(iii) Biochemical diagnosis

The object of biochemical investigation is to establish the existence and origin of cortisol excess.

(*a*) *Existence of cortisol excess*. The problem with most of the tests used is that biochemical values for patients with Cushing's syndrome overlap the normal range. Single tests are therefore of limited use and it may be that combining data from several tests in discriminant or quadric analysis can in-

Table 4. *Techniques for measuring cortisol*

Measurement of:	Principle	Other steroids measured	References
Urinary cortisol			
(1) 17-Hydroxysteroids	Colorimetry	11-Deoxycortisol	(242)
(2) 17-Oxogenic steroids (or 17-ketogenic steroids)	Colorimetry	Androgens	(243) (238)
(3) Free cortisol	Competitive protein binding	Corticosterone deoxycorticosterone 11-Deoxycortisol	(244)
(4) Cortisol secretion	Isotope dilution		(239)
Plasma cortisol			
(1) 11-Hydroxycorticosteroids	Fluorimetry	Corticosterone, oestrogens, 11-deoxycortisol, spironolactone; other non-steroidal substances	(240)
(2) 17-Hydroxycorticosteroids		As with urine	
(3) Cortisol	Double isotope derivative		(246)
(4) Cortisol	G.l.c. with electron capture		(245)
(5) Cortisol	Competitive protein binding using corticosterone-binding globulin	Corticosterone Deoxycorticosterone 11-Deoxycortisol	(241)

crease diagnostic precision. This is certainly the case in the differential diagnosis of primary hyperaldosterone (125).

Techniques used to measure cortisol are listed in Table 4. Cortisol excess, which characterizes the syndrome, is reflected in increased secretion of the hormone and increased urinary excretion of free cortisol, 17-hydroxycorticosteroids and 17-oxogenic steroids (52, 55, 57). Different laboratories favour different techniques but measurement of urinary free cortisol content is probably the most reliable (64, 65).

The plasma concentration of cortisol is raised and the amplitude of its normal circadian rhythm (maximum 6–8 a.m.; minimum 8–12 p.m.) is decreased. Thus plasma cortisol may be normal at 8 a.m., borderline at 8 p.m. and abnormally high at midnight (55). Measurement of the circadian rhythm of plasma cortisol, together with urinary free cortisol, are sufficient to establish diagnosis in most cases, but where difficulties arise an insulin tolerance test and measurement of cortisol secretion rate should be undertaken (52, 66).

Diagnostic problems may arise because of the increased 17-hydroxy- and 11-hydroxycorticosteroid excretion sometimes found in obesity (64, 65) and depression (67). Measurement of urinary free cortisol content is particularly useful in making the distinction (64, 65).

(*b*) *Differential diagnosis*. Having established that cortisol excess is present, the next problem is to determine its cause. Tests used in distinguishing the four main possibilities are shown in Table 5. Basal (8 a.m.) amounts of plasma ACTH are usually raised or in the upper part of the normal range in pituitary-dependent cortisol excess (52). As with cortisol, values at midnight are more often abnormal. ACTH is usually increased markedly in the ectopic ACTH syndrome (59) but because of the negative-feedback effect of increased cortisol (p. 11), plasma ACTH content is low in benign and malignant adrenal tumours (52).

Some discrimination is provided by the response of cortisol to injected ACTH (250 μg of tetracosactrin) and by large-dose (8 mg daily for 3 days) dexamethasone suppression (Table 5); but neither test can be relied on completely (64). Metyrapone (4.5 g over 24 h), by inhibiting 11β-hydroxylation (Fig. 4), decreases cortisol production and thereby stimulates endogenous ACTH secretion. This increases 17-oxogenic steroid excretion, and so distinguishes more clearly the pituitary-dependent disease from other types of Cushing's syndrome (Table 5, 52, 64).

(*iv*) *Hypertension in Cushing's syndrome*

The mechanism of hypertension in Cushing's syndrome is not certain. Sodium retention is present (55) and might be important (p. 35). Though most of the features of Cushing's syndrome can be attributed to cortisol excess (57) there are reasons to suspect that cortisol alone may not be responsible for the rise of blood pressure (1); hypertension produced by injected ACTH is more marked than that resulting from cortisol (68). Since plasma ACTH is often increased in Cushing's syndrome an alternative possibility (69) is that ACTH also increases deoxycorticosterone and corticosterone and that blood pressure is raised by these more pressor steroids.

(*v*) *Summary*

Cushing's syndrome is a dangerous, rare and incompletely understood cause of hypertension. Once considered, clinical diagnosis is often relatively simple, but biochemical confirmation is necessarily based on a number of investigations: measurement of urinary free cortisol concentration and the circadian rhythm of plasma cortisol to establish the presence of cortisol excess, and plasma ACTH and the metyrapone test to identify its cause.

7. Renin–Angiotensin–Aldosterone System

(*i*) *Renin and angiotensin*

Renin, a proteolytic enzyme (70, 71), is found mainly in granules within the wall of the afferent glomerular arteriole (72). Recent evidence suggests that it is mostly stored as an inactive proenzyme and that activation is associated with

Table 5. *Tests used to establish the cause of Cushing's syndrome*

Forms of Cushing's syndrome	Plasma ACTH	Cortisol response to ACTH	Plasma cortisol after 8 mg of dexamethasone	Response of 17-oxogenic steroids to metyrapone
(1) Pituitary-dependent adrenal hyperplasia	Increased or high normal	Usually increased	Suppressed	Marked increase
(2) Ectopic ACTH	Markedly increased	Usually normal	Less marked suppression	Variable changes
(3) Benign adrenal tumour	Low normal or unmeasurable	Variable	Less marked suppression	Slight decrease
(4) Malignant adrenal tumour	Low normal or unmeasurable	Usually negative	Less marked suppression	Variable changes

a decrease in molecular size (73). Active renin hydrolyses a leucyl-leucine linkage in a plasma protein substrate to produce a decapeptide, angiotensin I (Fig. 6). The same reaction occurs with the synthetic tetradecapeptide component of this substrate (74, Fig. 6).

Angiotensin I is probably biologically inactive and activation occurs when the *C*-terminal dipeptide is removed by so-called converting enzymes to produce angiotensin II. The process occurs in blood as it passes through the lungs (75, 76) and possibly to a lesser extent in other sites (75, 77, 78). Blood also has converting enzyme activity (79).

The octapeptide angiotensin II has a series of powerful pharmacological effects. On injection it raises blood pressure, stimulates aldosterone secretion, alters the urinary excretion of sodium and water and may stimulate release of antidiuretic hormone and catecholamines (see ref. 9). It provokes drinking in animals and may cause thirst in man (80).

A central question in work on renin and angiotensin is whether these pharmacological effects occur also as physiological or pathological phenomena (see ref. 9). Circulating concentrations of angiotensin II are within, or close to, a

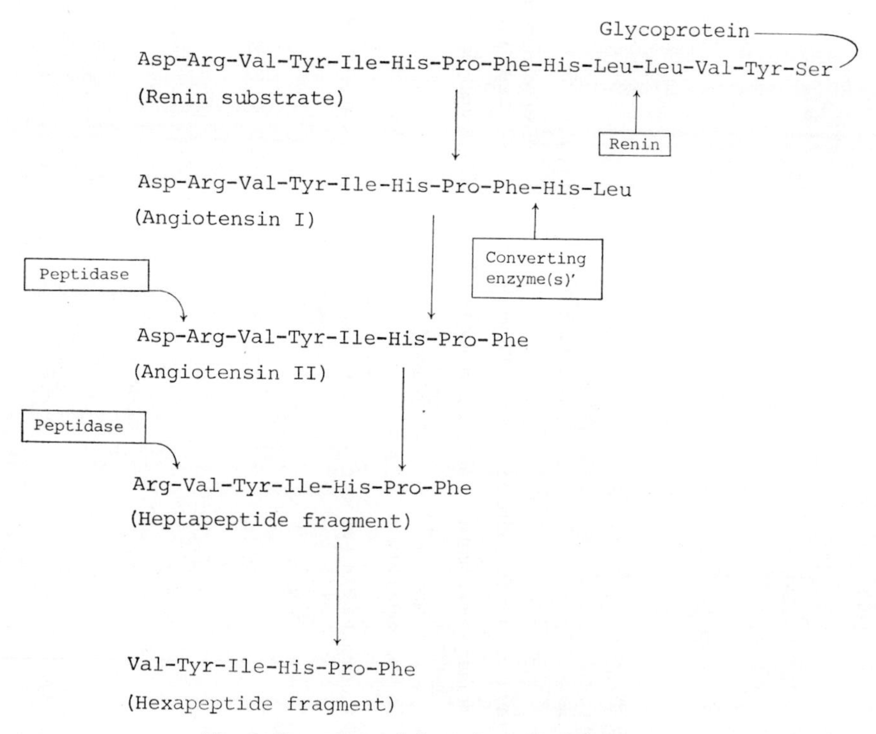

Fig. 6. *Components of the renin–angiotensin system*

range capable of affecting blood pressure (18), aldosterone secretion (82) and renal function (83). As is discussed below, sodium balance exerts an important moderating influence on two of these effects.

The plasma level of angiotensin II is controlled by a balance between enzymic production of the peptide (Fig. 6) and loss from the circulation by (*a*) diffusion or binding to receptors and (*b*) destruction by angiotensinases, a group of peptidases. The hepta- and hexa-peptide fragments produced by angiotensinase (Fig. 6) are important, partly because they cross-react with antisera raised against angiotensin II (78, 84), and partly because injections of the hepta-peptide stimulate aldosterone secretion (85).

(*ii*) *Stimuli of renin release*

This is a controversial field; measurement of renin release is technically difficult (86) and the number of stimuli producing release of renin is large

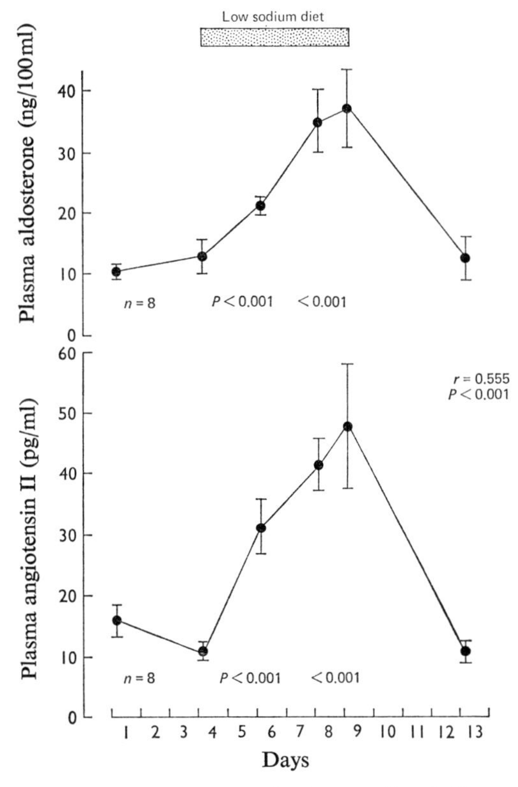

Fig. 7. *Increase in the plasma concentrations of angiotensin II and aldosterone in normal subjects during dietary sodium deprivation (less than* 10 *mequiv. of Na/day)*

(Reproduced by permission of *Lancet*)

(decreased artery pressure, loss of sodium or potassium, stimulation of renal nerves and injection of catecholamines and ACTH are a few). Evidence that there could be two main receptor mechanisms, (*a*) a pressure-sensitive afferent arteriole and (*b*) a sodium-sensitive macula densa, is discussed by Davis (87) and Thurau *et al.* (88). Whatever the exact mechanism involved, the main point to emerge is that sodium loss is usually associated with an increase in the circulating levels of renin and angiotensin II (71, 82, 89, 90).

(*iii*) *Control of aldosterone secretion*

Aldosterone is produced almost exclusively in the subcapsular zona glomerulosa of the adrenal cortex (Fig. 4). Injection or infusion of the steroid produces retention of sodium, loss of potassium and, after prolonged admini-

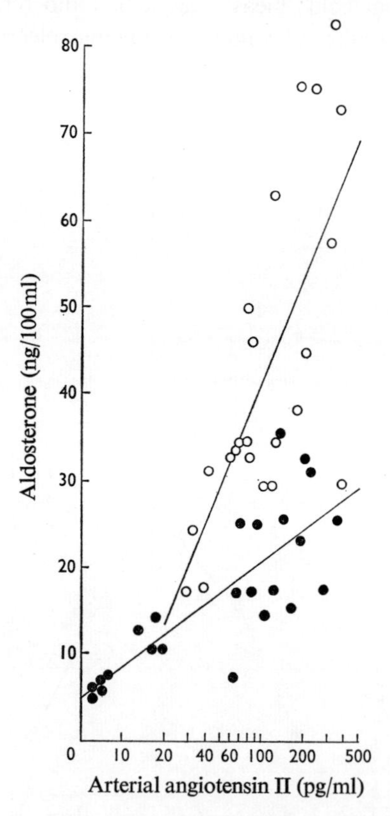

Fig. 8. *Correlation between plasma levels of angiotensin II and aldosterone in normal subjects, regression for data in the sodium-depleted state being steeper than that for sodium-replete subjects*

Upper points on both curves were obtained during infusion of angiotensin II. Data from Oelkers *et al.* (82). ●, Sodium-repleted subject; ○, sodium-depleted subject.

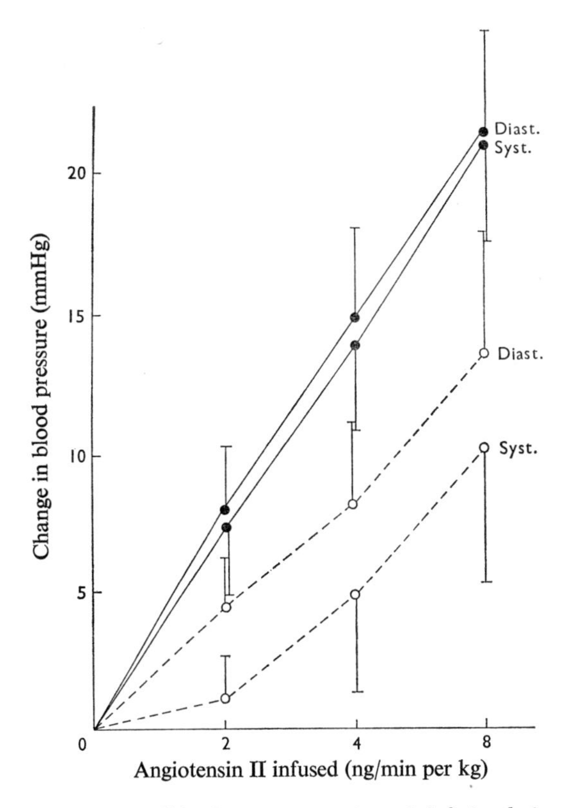

Fig. 9. *Decrease in response of blood pressure to angiotensin infusion during sodium deprivation in normal subjects*

Data from Oelkers *et al.* (82). Diast. = diastolic; Syst. = systolic. ●, Sodium-repleted subject; ○, sodium-depleted subject.

stration, an increase of blood pressure (91). Aldosterone influences sodium transport in a wide variety of biological membranes; Edelman & Fimognari (92) and Fraser (93) have reviewed evidence concerning the intracellular mechanisms which may be involved.

In normal man dietary sodium deprivation leads to a slight negative sodium balance and an increase in circulating concentrations of renin and angiotensin II (Fig. 7). Aldosterone concentration also rises partly because of the increased amount of angiotensin II and partly because sodium loss increases the response of aldosterone to angiotensin. This is seen by a steepening of the dose–response curve relating aldosterone concentration to angiotensin II concentration (Fig. 8). A second effect of sodium deprivation is to decrease the pressor response to angiotensin II (Fig. 9). The different changes of the two dose–response curves

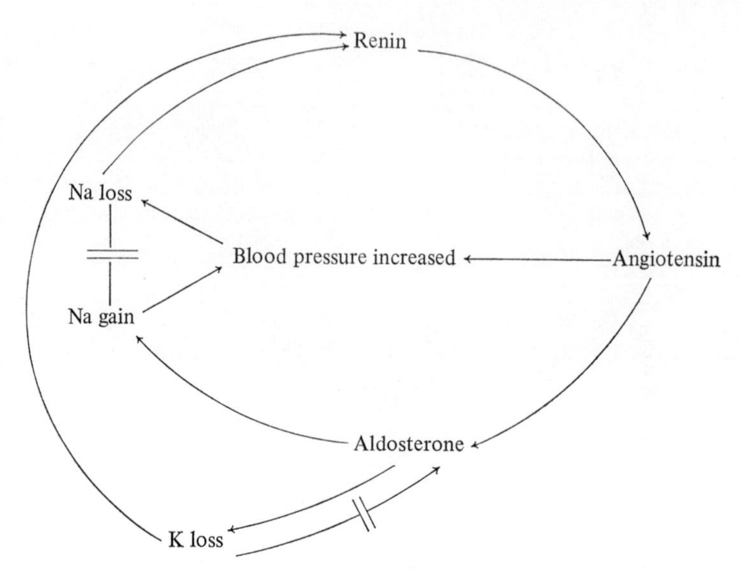

Fig. 10. *Hypothetical relation between renin, angiotensin, aldosterone and sodium balance*

on sodium deprivation both tend to diminish sodium loss and maintain homeostasis (Fig. 10, ref. 82).

Angiotensin II is not the only stimulus to aldosterone secretion. Increased plasma potassium and probably decreased plasma sodium have important effects (94–97). In some circumstances, ACTH also stimulates aldosterone (98–100). There has been much discussion on the relative importance of these four stimuli (82, 94, 101). Blair-West *et al.* (96) and Boyd *et al.* (102) suggest that there may be an as yet unidentified fifth stimulus.

Deriving from the views of Gross (103) Fig. 10 shows ways in which renin, angiotensin and aldosterone may regulate sodium balance. Sodium loss, by stimulating renin, angiotensin and aldosterone secretion, sets in motion a series of changes which restore sodium balance. Increased angiotensin will tend to raise blood pressure but the response of blood pressure to angiotensin would be decreased by sodium depletion (Fig. 8) and sodium loss would therefore be minimized. Potassium loss caused by aldosterone would tend to reduce aldosterone secretion and stimulate renin release.

Derangements of this mechanism are considered in more detail below.

(*iv*) *Techniques for the assay of renin, angiotensin and aldosterone*

The plasma concentration of renin can be measured by extracting the enzyme, incubating with exogenous substrate and measuring the rate of angiotensin formation using either bioassay (104) or the more recently developed radioimmunoassay (105). Renin activity is measured by incubating a plasma sample with endogenous substrate in the presence of inhibitors of angiotensinase. Renin then reacts with the variable amount of plasma substrate present to produce angiotensin. As with renin concentration techniques, radioimmunoassay has largely replaced bioassay. Commercial kits for the assay of plasma renin activity have been assessed recently (106). Renin activity is easier to measure but more difficult to interpret because of the varied contribution made by endogenous substrate (71, 84). The plasma concentration of renin substrate is determined by incubating plasma with excess of renin and measuring the angiotensin formed by bioassay (107) or radioimmunoassay (108).

The blood concentration of angiotensin I can be determined by extracting the peptide in organic solvents, followed by chromatography and radioimmunoassay (109). Angiotensin II concentration is measured more simply by chromatographic extraction followed by radioimmunoassay (78, 84). Plasma aldosterone concentration can be measured by double-isotope-derivative and radioimmunoassay techniques (110–112, 157). Methods for the measurement of aldosterone secretion and urinary excretion are discussed by Cope (66) and by Coghlan & Blair-West (111).

(*v*) *Relation between components of the renin–angiotensin system*

It would be expected, if the renin–angiotensin system operated *in vivo* as shown in Fig. 6, that plasma concentrations of renin and angiotensin II would

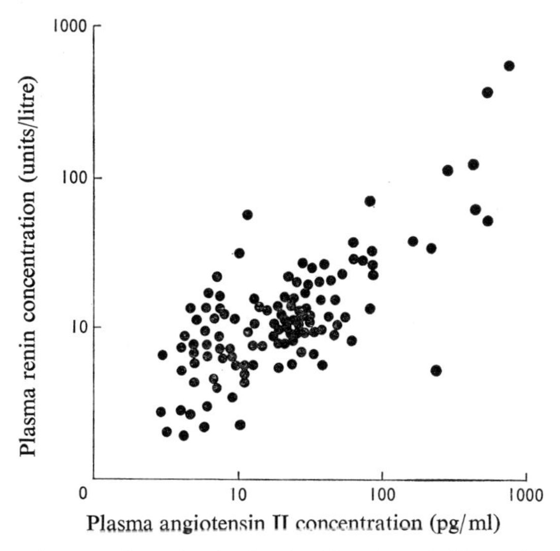

Fig. 11. *Relation between plasma levels of renin and angiotensin II in patients with hypertension of various types*

be related. Reasonably good correlations are indeed found in a wide variety of conditions and hypertension is no exception (Fig. 11). However, it is of some interest that the correlations in pregnancy and in women taking oral contraceptives are not significant ($n = 60$, $r = +0.24$, $P > 0.05$ for the former; $n = 31$, $r = -0.15$, $P > 0.05$ for the latter). There are several possible explanations for this; inactive renin may be released in greater amounts during pregnancy and activated during the measurement of renin concentration (113). Biologically inactive, but immunologically active, fragments of angiotensin II could also contribute to the poor correlation between renin and angiotensin II in this situation (78). Another possibility is that the high circulating concentration of renin substrate in pregnancy may alter the amount of angiotensin II formed by renin (114).

8. Hypertension with Aldosterone Excess

Aldosterone excess occurs in only a small number of patients with hypertension but these include several interesting and often treatable syndromes (69, 100, 115, 116). Hypokalaemia, usually the first indication of aldosterone excess, can have several explanations in a patient with hypertension.

(*i*) *Hypokalaemia with diuretic therapy*

Probably the commonest explanation for hypokalaemia is that the patient is receiving drugs of the benzothiadiazine (thiazide) group. The clinical importance of this is that it may mimic or obscure aldosterone excess from other causes. The mechanism of the hypokalaemia is not completely clear. Potassium loss induced directly by the diuretic acting upon the kidney tubule must play some part and it is possible that secondary hyperaldosteronism (see p. 28) induced by sodium loss also contributes (116–118). In patients with primary hyperaldosteronism the hypokalaemia produced by thiazide treatment may be particularly marked.

(*ii*) *Primary hyperaldosteronism; aldosterone excess with low plasma renin*

This syndrome is associated with an adrenocortical adenoma [Conn's syndrome (119)] or with bilateral hyperplastic and/or nodular change in the adrenal cortex. Very occasionally there are no detectable adrenal abnormalities. Symington (24) has given a full account of the pathological lesions.

(*a*) *Clinical and biochemical abnormalities.* Most features of the condition derive from aldosterone excess, and include potassium depletion, sodium retention, hypertension and depression of plasma renin concentration (Fig. 12). The hypertension is relatively mild and the malignant phase is rare but not unknown (3, 115). The majority of patients have no symptoms, though some may complain of weakness, nocturia and paraesthesiae, all of them manifestations

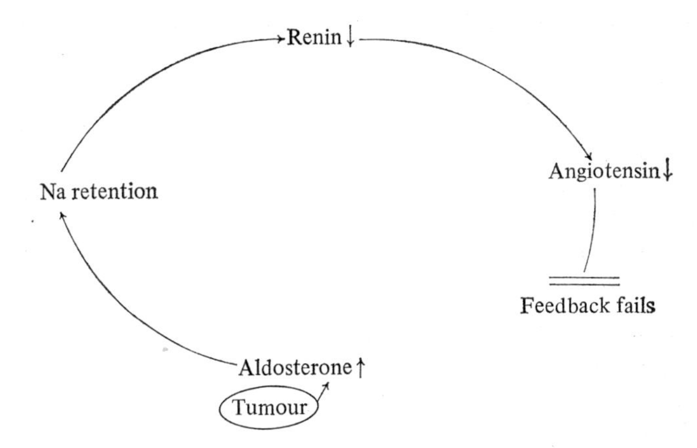

Fig. 12. *Oversecretion of aldosterone by a relatively autonomous adrenal lesion leads to sodium retention and hypertension with depression of renin and angiotensin*

of potassium depletion. Typically, the plasma potassium concentration is low but may be intermittently normal (120, 121). Plasma sodium concentration is raised or in the upper half of the normal range. This is an important diagnostic distinction from secondary hyperaldosteronism in which plasma sodium concentration is low or low normal.

Extracellular alkalosis in primary hyperaldosteronism is reflected in a raised plasma bicarbonate content. Total exchangeable sodium content, total body water and extracellular fluid volumes are increased while exchangeable total body potassium content is decreased (Fig. 15 and refs. 100, 115, 122, 123). Diagnosis is based on the association of aldosterone excess with diminished plasma renin concentration or renin activity. Aldosterone excess is demonstrated by measuring the secretion (66), urinary excretion (124) or plasma concentration (120) of the hormone.

Pre-operative differentiation of patients with and without adrenal tumour is important as operation is indicated for the former, and prolonged treatment with spironolactone (see below) may be preferable for the latter. Biochemical tests viewed singly are of no help in this but quadric analysis, using a combination of biochemical and clinical data for individual patients, makes a clear distinction (125).

(*b*) *Treatment of primary hyperaldosteronism.* Blood pressure is usually controlled, and the electrolyte abnormality is invariable corrected, by spironolactone although relatively large doses (300–400 mg/day) may be needed (116, 123, 126). Side effects are probably less marked than with other potent hypotensive agents, but when they occur amiloride, another potassium-sparing diuretic, is a useful alternative (127).

Unilateral adrenalectomy is indicated in patients with an adenoma and the results are generally good (128), comparable in individual patients with their previous response to spironolactone (116). The drug is therefore useful in predicting the outcome of surgery.

The ideal treatment for patients with bilateral hyperplasia or nodular change in the suprarenal is not established (116, 124, 128). Total or sub-total bilateral adrenalectomy is often effective but the operation is more liable to complications than unilateral adrenalectomy and life-long corticosteroid replacement therapy may be needed subsequently. The alternative of prolonged treatment with spironolactone could be preferable.

(*iii*) *Intermittent primary hyperaldosteronism*

As with other corticosteroids, plasma levels of aldosterone show distinct circadian changes both in normal subjects and in patients with primary hyperaldosteronism (129, 130). In our experience plasma aldosterone assays have generally given unequivocal results with samples taken between 8 and 10 a.m. although in an earlier study (120) the plasma aldosterone concentration was raised only intermittently in some patients. We have seen this since in 8 of 31 (26 %) patients with histologically proven primary hyperaldosteronism. In most of these the difference between normal and abnormal values was not large. A more difficult diagnostic problem was present by the case described below:

> The patient, a woman of 59, presented in 1971 with severe weakness, hypertension (200/110) and a plasma electrolyte pattern typical of aldosterone excess (Na, 145; K, 2.8; HCO_3, 36 mequiv./l). Subsequently, on two occasions plasma aldosterone was normal (checked by g.l.c. and radioimmunoassay) in the presence of a high plasma sodium, a low plasma potassium and depression of renin (Fig. 13) and angiotensin II (5–7 pg/ml). Plasma levels of deoxycorticosterone (4.1 ng/100 ml) and cortisol (7.3 μg/100 ml) were also normal. Later, and without introduction of treatment, aldosterone increased to grossly abnormal values (Fig. 13). Adrenal venography demonstrated a tumour in the left adrenal gland and adrenal venous plasma from this side had an aldosterone concentration of 1760 ng/100 ml as compared with 88 ng/100 ml from the right side. Subsequently, a 20 mm tumour was excised and its histology was typical of an aldosterone-secreting adenoma. Blood pressure, plasma electrolytes, renin, angiotensin and aldosterone have been normal since operation.

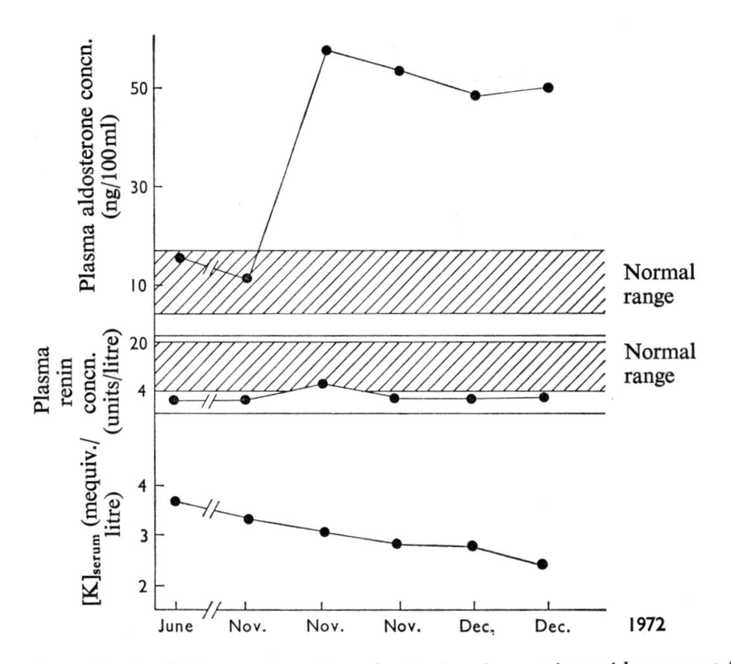

Fig. 13. *Plasma levels of aldosterone, renin and potassium in a patient with apparent intermittent primary hyperaldosteronism*

Except for the sample in June 1972, when the patient was on 100mg of spironolactone daily (the effect of spironolactone is to raise renin, potassium and aldosterone), all measurements were made in the untreated state, therapy having been withheld for at least 4 weeks.

So far as we know, there have been no previous reports of intermittent primary hyperaldosteronism. Cortisol is secreted intermittently in normal man and intermittent Cushing's syndrome has been described (131). The practical point raised by the case described above is that two normal values of plasma aldosterone cannot exclude otherwise typical severe primary hyperaldosteronism.

(iv) Possible association of primary hyperaldosteronism and a renal lesion

Primary hyperaldosteronism and renal abnormalities are rare causes of hypertension. Though it may be a coincidence we have, among 130 cases of primary hyperaldosteronism, seen 15 patients with an additional renal abnormality (Table 6). Gowenlock & Wrong (132) and Bloch (133) have described similar cases.

As described earlier angiotensin probably acts as a trophic hormone for aldosterone and prolonged stimulation of the aldosterone biosynthetic mechanism might lead to irreversible hyperplasia or tumour formation (132, 134). A similar process has been invoked to explain tertiary hyperparathyroidism and

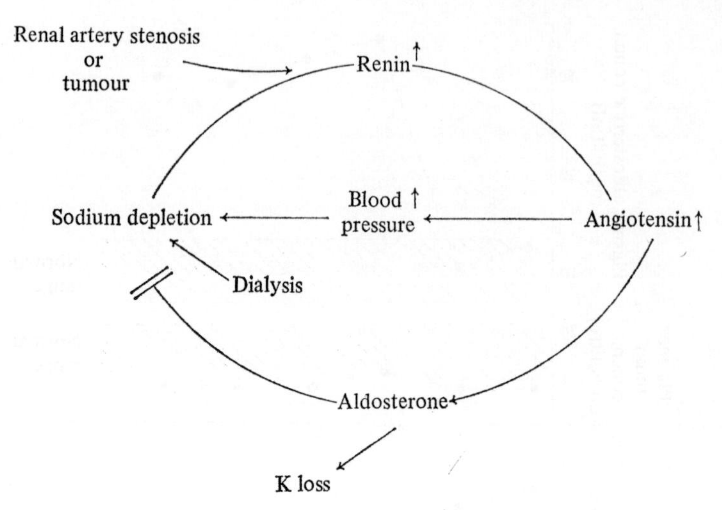

Fig. 14. *Secondary hyperaldosteronism*

it is of interest that patients with pituitary-dependent Cushing's syndrome also sometimes have tumours of both the pituitary gland and adrenal cortex (60).

(v) Hyperaldosteronism suppressible by dexamethazone

A small number of patients have hyperaldosteronism with low renin and hypertension yet in whom the syndrome is suppressible by dexamethazone (135–137).

(vi) Secondary hyperaldosteronism in hypertension

This condition develops when aldosterone is stimulated by an excess of renin and angiotensin.* Plasma sodium concentration is often decreased. Fig. 14 shows one way in which the abnormalities may be related. Plasma renin concentration is increased either because renin release is abnormally high in relation to sodium balance (a tumour or renal artery stenosis) or because hypertension, or treatment of hypertension by dialysis (see below), has produced sodium depletion.

These features contrast with primary hyperaldosteronism (Fig. 12) where renin and angiotensin are low and plasma sodium is increased. The distinction is important because treatment of the two syndromes is quite different; mistakes occur and the adrenal glands are sometimes wrongly excised in secondary hyperaldosteronism before the renal cause of the adrenal abnormality becomes apparent. Differential diagnosis can be made, quite simply, by measurement of renin and/or angiotensin concentrations. Secondary hyperaldosteronism is

* Strictly the same term should also be applied to over-stimulation of aldosterone by other agents, but clinical examples of these have yet to be described.

Table 6. *Primary hyperaldosteronism*

Fifteen patients with primary hyperaldosteronism associated with a renal lesion are shown. R, right; L, left; RAS, renal artery stenosis; IVP, intravenous pyelogram; FMH, fibromuscular hyperplasia.

Patient			Adrenal lesion		Renal lesion	
No.	Sex	Age	Nature	Evidence	Nature	Evidence
1	F	36	Adenoma (L)	Biochemistry, pathology and surgery	R. hydronephrosis and renal failure	IVP
2	M	49	Adenoma (L)	Biochemistry, pathology and surgery	R. RAS and FMH	Renal arteriogram
3	F	50	Adenoma (R)	Biochemistry, pathology and surgery	Bilateral RAS	Arteriogram and operation
4	F	50	Adenoma (L)	Biochemistry, pathology and surgery	Bilateral hydronephrosis	IVP
5	M	33	Adenoma (L)	Biochemistry, pathology and surgery	Medullary sponge kidney	IVP
6	F	50	Adenoma (L)	Post mortem	RAS (R)	Post mortem
7	M	49	Adenoma (L)	Biochemistry, pathology and surgery	Bilateral hydronephrosis	IVP
8	M	64	Hyperplasia	Biochemistry, pathology and surgery	Carcinoma of L. kidney	Operation
9	F	41	Hyperplasia	Biochemistry, pathology and surgery	RAS (L)	Arteriogram and isotope renogram
10	F	59	Adenoma	Biochemistry and quadric analysis	R. hydronephrosis and stone	IVP
11	F	56	Adenoma	Biochemistry and quadric analysis	Polycystic kidney	IVP
12	F	46	Adenoma	Biochemistry and quadric analysis	RAS (R)	Arteriogram and operation
13	M	58	Adenoma	Biochemistry and quadric analysis	RAS (R)	Arteriogram
14	F	65	Adenoma	Biochemistry and quadric analysis	RAS (R)	IVP+divided studies
15	F	45	Hyperplasia	Biochemistry and quadric analysis	Bilateral RAS	Arteriogram

found with malignant-phase hypertension, renal hypertension [particularly renal artery stenosis (138)] the intractable hypertensive syndrome which develops in chronic renal failure (139) and renin-secreting tumour (140). These are discussed in more detail on page 32.

Secondary hyperaldosteronism also occurs in several conditions associated with normal or low blood pressure including pregnancy, cirrhosis, sodium-losing renal disease and anorexia nervosa (89, 100).

Management of secondary hyperaldosteronism may be more difficult than primary hyperaldosteronism. Anorexia and wasting are more common and there is sometimes thirst, polydipsia and polyuria. Hypertension is usually severe and sometimes difficult to control by drugs. Again, in contrast with primary hyperaldosteronism, spironolactone is generally ineffective in controlling blood pressure. However, the syndrome does respond to a decrease in blood pressure whether this is achieved by drugs, renal vascular surgery or unilateral nephrectomy (100, 141, 142). As will be described, bilateral nephrectomy may be justified in cases with intractable hypertension and advanced renal failure (139). In each of these instances successful control of the disorder is associated with a decrease of renin and angiotensin to normal or subnormal levels.

(*vii*) *Summary*

Hypertension with excess aldosterone secretion usually occurs with hypokalaemia. It occurs in two main forms: in primary hyperaldosteronism, excess of the steroid derives from a tumourous or hyperplastic adrenal cortex. Diagnosis is based on the combination of aldosterone excess with sodium retention and decreased plasma renin. Spironolactone usually controls blood pressure and always controls the electrolyte abnormality. Surgery, if needed, is directed at the adrenal gland and is usually successful.

In secondary hyperaldosteronism hypertension is more severe and does not usually respond to spironolactone. Surgery, if needed, is directed at the kidney and is usually curative.

9. Hypertension with Excess of Other Mineralocorticoids

(*i*) *Deoxycorticosterone*

The ability of injected deoxycorticosterone acetate to raise blood pressure in animals is well known (143, 144). Only recently, however, have techniques of sufficient sensitivity been developed to measure the concentration of deoxycorticosterone in plasma (112, 145) since which time six hypertensive patients with apparently isolated excess of deoxycorticosterone have been reported (146). Aldosterone levels in these were generally normal and their hypertension and electrolyte abnormality responded well to spironolactone. Micronodular hyperplastic changes in the adrenal cortex have since been found in one of these patients (127).

(*ii*) 18-*Hydroxydeoxycorticosterone*

Another candidate for the production of mineralocorticoid hypertension is 18-hydroxydeoxycorticosterone. Hypertensive patients with high urinary excretion of this steroid have been reported recently (147, 148).

(*iii*) *Corticosterone*

Corticosterone is a further possibility but, except for one patient with corticosterone-secreting carcinoma of the adrenal cortex (149) and congenital adrenal hyperplasia (see below) there have been few reports.

Lack of opportunity to apply the new analytical techniques for steroids in biological fluids to an unselected hypertensive population make it impossible to estimate the prevalence of excess of deoxycorticosterone, 18-hydroxy-deoxycorticosterone and corticosterone in hypertension.

(*iv*) *Congenital adrenal hyperplasia with* 17α-*hydroxylase deficiency*

This is a rare but interesting cause of hypertension usually presenting in children. Hereditary deficiency of the adrenal enzyme 17α-hydroxylase blocks cortisol synthesis (Fig. 4) and diverts precursor progesterone into excess production of corticosterone and deoxycorticosterone (124, 150, 151) which are probably responsible for the increased blood pressure. Oestrogen synthesis is also blocked and female patients show hypogonadism in contrast with the 11-hydroxylation deficiency syndrome in which virilism is a feature (see below). Because aldosterone production is also sometimes impaired there may be additionally a deficiency of the enzymes responsible for the conversion of corticosterone into aldosterone (150, 151). Alternatively, decreased renin production may be responsible for the observed decrease in aldosterone (152).

(*v*) *Congenital hyperplasia with* 11-*hydroxylation deficiency*

This is a further cause of hypertension in children. Deficiency of 11-hydroxylation is associated with a decrease in cortisol, corticosterone and aldosterone production and the hypertension is thought to be due to the marked increase of deoxycorticosterone (124, 153). In contrast with the 17α-hydroxylation syndrome, children with the 11-hydroxylation disorder usually present with virilization.

10. Hypertension with Liquorice and Carbenoxolone

Hypertension with hypokalaemia and sodium retention has been reported in patients taking a large quantity of liquorice (154, 155). A similar syndrome sometimes develops during treatment with carbenoxolone (156, 158).

11. Renal Hypertension

Hypertension can produce renal vascular abnormalities and renal vascular abnormalities can produce hypertension (159, 160) so that distinguishing cause and effect can sometimes be difficult in patients with renal hypertension. Further, although disease of one kidney can raise blood pressure (particularly in man and the rat), removal of that kidney does not always decrease blood pressure. When operation fails, changes produced by hypertension in the contralateral (untouched) kidney may be responsible (160–163).

The main problem with renal hypertension in man is not diagnosis but identification of patients likely to respond to surgical correction of the renal abnormality. Among the commoner causes of hypertension are stenosis and thrombosis of the renal artery, polycystic kidney, acute and chronic glomerulonephritis, renal tuberculosis, renal tumours (particularly renin-secreting tumours), polyarteritis nodosa and systemic lupus erythematosis. Regardless of its cause chronic renal failure is often associated with hypertension. Pickering (1) and Kaplan (3) have dealt with these and other causes of renal hypertension in greater detail.

Tests used in diagnosis include the intravenous urogram, renal arteriography, isotope renography, bilateral ureteric catheterization studies and measurement of renin or renin activity in the peripheral and renal venous plasma. The relative merits of these have been fully discussed elsewhere (9, 142, 164, 165).

Renin is stored in large amounts in the kidney and in even larger amounts in kidneys with renal artery stenosis. Not surprisingly, during the 1950s and 1960s, before methods for measuring plasma renin concentration became available, it was widely believed that renal hypertension was caused directly by excess of circulating renin. In fact, this is not the case; hypertensive patients with undoubted renal artery stenosis may have normal circulating concentrations of renin (71) and in the dog only the early stages of experimental renal hypertension are associated with increased renin (166). As is discussed in the second part of this essay, an abnormal relation between renin and sodium may be more important in renal hypertension than an isolated increase of renin.

(*i*) *Renal artery stenosis*

Stenosis and thrombosis of the renal artery is one of the commoner causes of renal hypertension in man. The hypertension is often severe and sometimes in the malignant-phase with retinal haemorrhages and exudates (1, 3, 164, 167). Secondary hyperaldosteronism is sometimes present (138) and, when severe, may be associated with gross depletion of water, sodium and potassium (168).

In the absence of secondary hyperaldosteronism biochemical tests are of little help in differentiating renal from other causes of hypertension, although a raised renin concentration or renin activity in venous plasma from an affected kidney is a favourable omen for surgery (3, 142, 169).

(ii) Renin-secreting tumour

Seven cases of renin-secreting haemangiopericytoma (juxtaglomerular cell tumour) of the kidney have been reported in the last few years (140, 170–172). Renal tumours of other types also occasionally secrete an excess of renin (140). The syndrome is of great interest because it provides probably the best evidence that excess of renin can raise blood pressure and stimulate aldosterone secretion. Generally, patients are young and their hypertension severe. Hypokalaemia has been present in all of the cases so far reported and, where measured, aldosterone has also been increased. Usually, but not always, a tumour has been seen or suspected on arteriography.

Probably the best diagnostic test is the measurement of renin in samples of renal venous and lower inferior vena cava plasma taken simultaneously (inferior vena cava and arterial renin values are comparable). The differences found in the two renal veins are small. This is not surprising since, provided renal blood flow is not decreased, the kidneys of man and the dog maintain a wide range of peripheral renin values with a veno–arterial renin difference of only 25% (86, 173). Thus, when one kidney only is releasing renin, as is usually the case with a renin-secreting tumour, the veno–arterial difference may rise to 50% but differences larger than this result from decreased renal plasma flow. Though diagnosis is difficult it is clearly worthwhile, because in each case reported so far, surgical removal of the tumour has relieved both the hypertension and secondary hyperaldosteronism.

Difficulties in diagnosing this syndrome emphasize again the differences between primary and secondary hyperaldosteronism (Figs. 12 and 14). Two of the seven patients with renin-secreting tumours were subjected to adrenal surgery first because primary hyperaldosteronism was suspected. Only later was attention diverted to the kidney. The diagnosis should be considered in any hypertensive patient whose persistently increased renin is not attributable to therapy, malignant-phase hypertension or a renal vascular lesion.

(iii) Hypertension with chronic renal failure

Most patients with chronic renal failure have increased blood pressure by the time regular dialysis treatment is needed. The mechanism of the syndrome is of great interest and is discussed in more detail in the second part of this essay. Abnormalities found in the majority of patients are increased total exchangeable sodium and total body water with expansion of extracellular and plasma volumes (139, 174). Plasma renin and angiotensin concentrations are usually close to the normal range (139). Regular dialysis treatment usually controls blood pressure and it seems likely that it does so by decreasing the excess of sodium and water (139, 174), although hypotensive drug therapy may have to be continued.

In a much smaller number of patients (the proportion varies markedly from one centre to another) blood pressure cannot be controlled by dialysis and may actually increase during the unsuccessful attempt; periods of severe hypertension alternating with postural hypotension (175). Usually secondary hyperaldosteronism has developed at this stage and sodium and water depletion (attributable to dialysis) is associated with markedly increased circulating concentrations of renin, angiotensin II and aldosterone (139, 176). Excruciating thirst is a serious problem in some of these patients (all necessarily being on restricted fluid intake) and it may be that increased angiotensin II is responsible (80, 139).

Bilateral nephrectomy can control most of these problems. After operation, the blood pressure decreases to normal and becomes much easier to control by dialysis; plasma concentrations of renin, angiotensin I, angiotensin II and aldosterone decrease to normal or subnormal after operation, the fall in renin being related to the fall of blood pressure (176). The patient's general condition improves and lost weight is regained. However, the anaemia, which is often present before operation, usually becomes more marked. Nevertheless, with renal transplantation as a long-term prospect, regular dialysis need not necessarily be life-long.

12. Malignant-Phase Hypertension

With the possible exception of coarctation, malignant hypertension may develop in patients with high blood pressure of almost any aetiology. Without treatment, expectation of life is less than one year (1).

Clinical features of the malignant hypertension syndrome include a rising blood pressure, fits, retinal haemorrhages and exudates, papilloedema, proteinuria and renal failure. The pathological hallmark is a fibrinoid lesion of arterioles particularly those in the kidney (177). Secondary hyperaldosteronism with increased renin and aldosterone are also present in some cases (see above). The mechanism of the disorder is unclear, though it is likely that increased arterial pressure plays a primary role (1). One possibility is that raised arterial pressure increases the permeability of blood vessel walls; plasma proteins including fibrinogen, penetrate (177) and coagulation develops within the walls and lumen of the vessel. It has been suggested (178), that this is part of the syndrome of disseminated intravascular coagulation with microangiopathic haemolytic anaemia. There is certainly evidence of both in patients and animals with malignant-phase hypertension (178), and it may be that they play a part in the transition from benign to malignant hypertension.

13. Hypertension on Oestrogen/Progestagen Oral Contraceptives

Blood pressure rises slightly but distinctly in most women taking oestrogen/progestagen oral contraceptives (179, 180). Severe hypertension, although rare, has also been reported in these circumstances (181, 182). Blood pressure may also rise still further when a hypertensive woman takes oral contraceptives (181). In most of these situations blood pressure falls when the contraceptive is stopped (180–182). The mechanism of the hypertensive effect is uncertain. Oral contraceptives provoke a large increase in plasma renin-substrate concentration (180, 181, 183), a compensatory fall in renin concentration and a variable change in angiotensin II. Renin activity is increased (183, 184).

14. Essential Hypertension

With the provisos mentioned earlier a diagnosis of essential hypertension would be made in a patient with a diastolic pressure persistently above 100 mmHg and in whom no clinical, biochemical or radiological evidence of secondary hypertension can be found (Table 2). The frequency with which the diagnosis of essential hypertension is made depends on the tests used. Clearly, essential hypertension would be diagnosed less often if intravenous urography, renal arteriography and isotope venography were done on all hypertensives together with measurement of plasma electrolytes (on several occasions), renin, angiotensin, urinary vanillylmandelic acid, metadrenalines and free catecholamines; urinary and plasma corticosteroids, including aldosterone, 11-deoxycorticosterone, corticosterone and even 18-hydroxydeoxycorticosterone. In practice no clinician would press diagnosis this far. With rare exceptions (see below), there is usually no justification for renal arteriography in a patient with a normal intravenous urogram and there may be less need than was formerly thought for a urogram if other evidence of renal disease is lacking (185). Urinary vanillylmandelic acid is rarely assayed unless clinical suspicion of phaeochromocytoma is aroused. 18-Hydroxydeoxycorticosterone, deoxycorticosterone, aldosterone and renin can be assayed only in a few centres and, even in these, measurements would only be made if other evidence of mineralocorticoid excess (such as hypokalaemia) were present.

A workable compromise would be to arrange a chest X-ray, intravenous urogram, estimation of plasma electrolytes, blood urea and serum creatinine, blood count and little else on a newly diagnosed hypertensive patient. Investigation would only be pressed further if these proved abnormal, if the patient were young or if the hypertension were severe or unresponsive to therapy. However, most specialist centres settle for a programme of investigation more than this and less than the theoretical maximum described above. For this reason the criteria for diagnosis of essential hypertension will vary from one centre to another.

Several attempts have been made to subdivide essential hypertension. As is discussed below, low-renin hypertension is considered by some to represent a distinct syndrome.

In summary, essential hypertension is an arbitrary diagnosis made by a process of exclusion. Biochemical tests play a large part in this. Essential hypertension is important because it is common and liable to serious complications. Treatment can certainly decrease blood pressure and, when diastolic levels are initially greater than 109 mmHg, also it reduces the risk of complications.

PART II

BIOCHEMICAL ABNORMALITIES IN THE PATHOGENESIS OF HYPERTENSION

As described earlier arterial pressure is controlled by a number of mechanisms. Psychological processes, the central and peripheral nervous system, the heart and blood vessels, sodium and water, catecholamines, the renin–angiotensin–aldosterone system and probably others play a part. Emphasis in this essay is on biochemical mechanisms and the reader should consult other publications for information on psychological and nervous factors (186, 187, 211), and the role of the heart and blood vessels (188, 189). Guyton *et al.* (190) have given an interesting account of ways in which the various mechanisms may interrelate; primary importance in the long-term regulation of blood pressure is attached to the process by which increased arterial pressure leads to loss of sodium and water from the kidney.

1. Sodium and Water in Hypertension

It is well known that low-salt diets can decrease blood pressure in man and the success of treatment by diuretic agents may be a modern counterpart. Experimentally, it has been shown that increased sodium intake can raise the blood pressure in animals, particularly those with genetic susceptibility (191) or impaired renal function (192). The ability of sodium-retaining steroids, particularly deoxycorticosterone, to raise the blood pressure is also established though again the effect is more marked when combined with high sodium intake and more marked still when kidney tissue is removed (143, 144, 193). Exchangeable body sodium concentration and blood pressure correlate significantly in rats with hypertension induced by deoxycorticosterone (193).

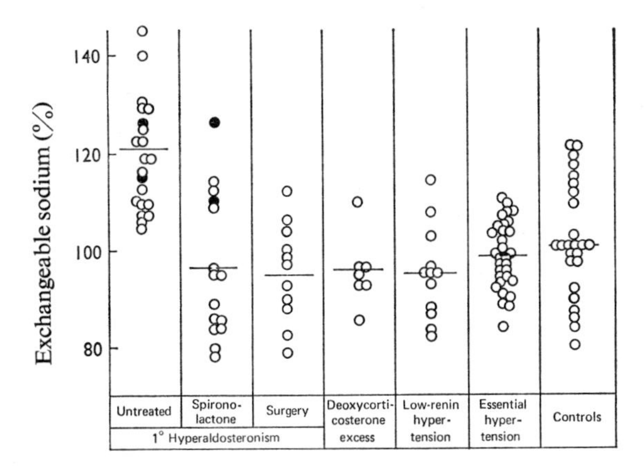

Fig. 15. *Exchangeable sodium, expressed as a percentage of the value expected for a normal subject of the same leanness index* [*Davies et al.* (203)] *in primary hyperaldosteronism, hypertension with apparently isolated deoxycorticosterone excess, low-renin hypertension and essential hypertension*

Only the mean value for untreated primary hyperaldosteronism differs significantly from normal ($t = 5.11$, $p < 0.001$]. Among cases with primary hyperaldosteronism (○) represents those with and (●) those without an adrenal adenoma.

Despite this suggestive evidence, patients with essential hypertension do not have increased plasma or exchangeable body sodium concentrations (Fig. 15), although they may show a correlation between the exchangeable sodium concentration (within the normal range) and blood pressure (194). Plasma and ECF volumes are also usually normal (195, 247). In contrast, patients with secondary hyperaldosteronism are sometimes sodium-depleted (p. 28).

As a further contrast the exchangeable sodium concentration is increased in patients with primary hyperaldosteronism (Fig. 15 and refs. 122, 123, 196, 206). Treatment with spironolactone decreases the exchangeable sodium concentration and blood pressure to normal (122, 196, 206). The exchangeable sodium and blood pressure are also closely related in hypertensive patients with chronic renal failure (197–199). Here again, removal of sodium and water (by dialysis) decreases blood pressure (139, 174, 196).

There is therefore one hypertensive syndrome with an excess of body sodium and a positive correlation of body sodium content and blood pressure, another with a normal body sodium content and a third with subnormal body sodium content and a negative correlation of body sodium content and blood pressure (194). This indicates that, with the exception of primary hyperaldosteronism, hypertension is not maintained, in the steady state, by an excess of sodium. It does not, however, exclude a role for sodium excess in the earlier stages of the disease (9).

Borst & Borst-de Geus (154) noted in 1963 that sodium retention might occur initially in the hypertensive syndrome associated with liquorice ingestion and that subsequently, as a result probably of the natriuretic effect of increased blood pressure, sodium excess was eliminated and sodium balance restored. They, and others since (see refs. 9, 248), suggested that the pressure–natriuretic mechanism may be reset at a higher level in hypertension. There is recent experimental evidence for this (200). As discussed below, the resetting process could be a cause or a consequence of increased blood pressure.

2. Renin and Blood Pressure

Evidence connecting renin and blood pressure is equally difficult to interpret. On the one hand it seems very likely that the amount of renin and angiotensin II in blood is within, or close to, a range capable of influencing blood pressure (81, 201), while on the other the relation between blood pressure and the plasma levels of renin and angiotensin can be significantly positive, non-existent or significantly negative (see ref. 194). It seems likely therefore that some factor or factors override the influence of renin on blood pressure or that the pressor response to renin is altered. There is good evidence that changes of sodium balance can act in this way (9, 82, 202 and Fig. 9). The possible importance of this effect is discussed in the following section.

As described above excess of renin is probably responsible for the hypertension and excess of aldosterone in patients with renin-secreting tumour and the intractable hypertensive syndrome in chronic renal failure, but these are the only syndromes where the relation is at all clear. In renal artery stenosis, renin is usually higher than normal, but not invariably so, and the correlation of renin and angiotensin with blood pressure in renal and malignant hypertension, though significant, is usually poor (194).

Thus, excess of sodium and excess of renin, alone, account for only a small number of hypertensive patients. The section which follows is concerned with the possibility that an abnormal relation between renin and sodium may be more important.

3. Relation of Renin and Sodium in Hypertension

Normally, sodium and renin concentrations are inversely related; loss of sodium leads to an increase of renin in the plasma and sodium retention suppresses renin secretion (Figs. 7 and 10). Because excess of sodium and excess of renin are both potentially pressor an inverse relation between the two mechanisms will tend to stabilize blood pressure since overactivity of one would be compensated by underactivity of the other.

(i) Renin concentration inappropriately high in relation to exchangeable sodium concentration

It has been suggested that failure of this compensatory mechanism may be responsible for raising blood pressure in renal hypertension and that renin may be excessively high in relation to sodium status (9, 174). Fig. 16 shows the inverse relation between exchangeable body sodium and plasma renin concentration in patients with normal blood pressure. As compared with this, hypertensive patients with chronic renal failure or malignant-phase hypertension have circulating concentrations of renin which are usually abnormally high in relation to concurrent exchangeable sodium (Fig. 17). In patients whose

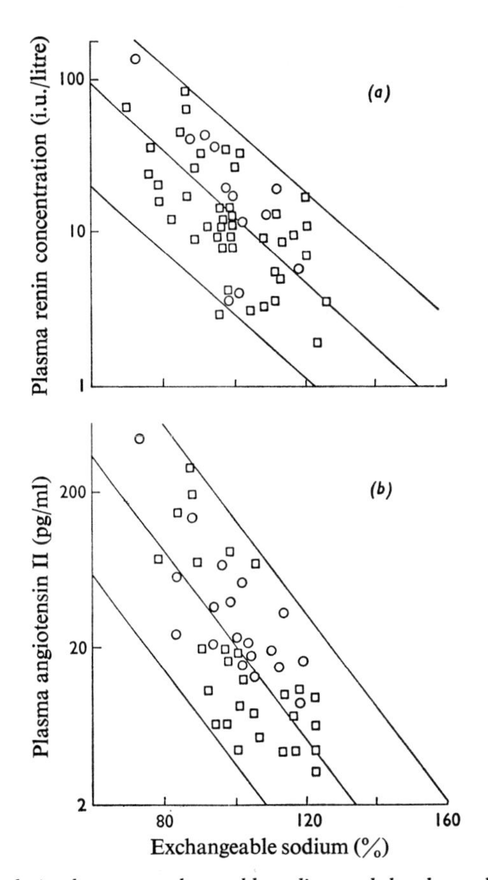

Fig. 16. *Inverse relation between exchangeable sodium and the plasma levels of renin* (*a*) *and angiotensin II* (*b*) *in subjects with normal blood pressure*

○, Patients with renal failure; □, other conditions (normal subjects, peptic ulcer, thyroid disease and Addison's disease). From Davies *et al.* (203). (Reproduced by permission of *Lancet.*)

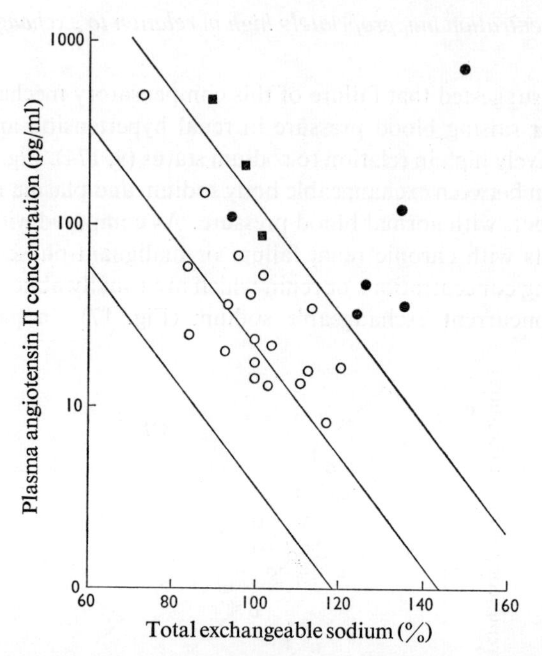

Fig. 17. *Deviation from the normal exchangeable sodium/angiotensin relation (lines indicate mean ±2* s.d. *from Fig. 16) in hypertensive patients with chronic renal failure (●) and malignant-phase hypertension (■)*

Normotensive patients with chronic renal failure are shown for comparison (○). From Davies *et al.* (203). (Reproduced by permission of *Lancet*.)

blood pressure is subsequently controlled by sodium-depleting regular dialysis treatment, the relation of exchangeable sodium and renin becomes normal (196, 203). Failure of renin to rise in the normal way on sodium depletion may be the reason for this therapeutic success.

As described earlier, hypertension occasionally follows a more intractable course in chronic renal failure. Instead of decreasing during dialysis, blood pressure may actually rise and the attempt to control it by sodium depletion is associated with progressive secondary hyperaldosteronism (Fig. 14). Circulating levels of renin, angiotensin I and angiotensin II are much higher in these patients (139, 176) and there is evidence that they rise during and as a result of the unsuccessful attempt to control blood pressure (Fig. 18). This contrasts with the relatively stable renin level in controllable cases (Fig. 18) and while sodium depletion is probably less marked in these, there could be a more important basic difference in the two syndromes. Fig. 19 illustrates this; it is suggested that in the intractable hypertensive state, renin is abnormal in relation to sodium status, but that the renin release mechanism is responsive to sodium depletion,

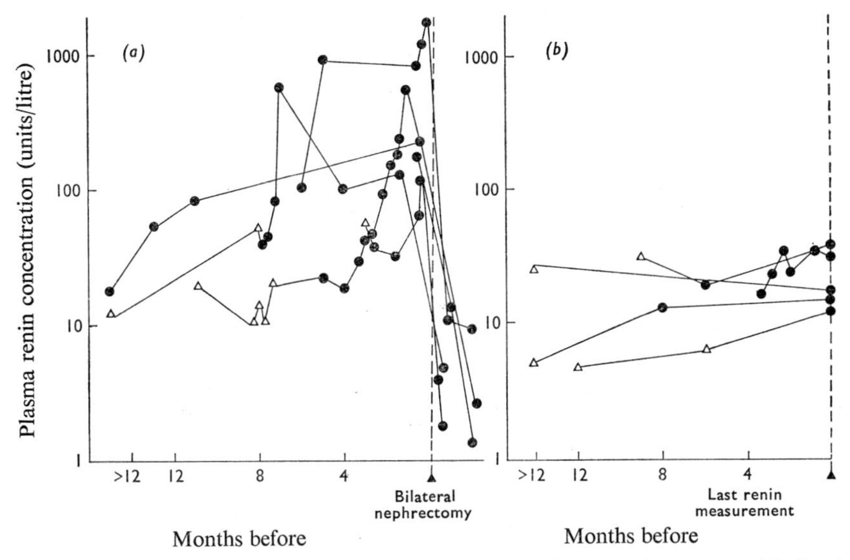

Fig. 18. *Serial measurements of plasma renin concentration in hypertensive patients with chronic renal failure*

Renin in the five patients with intractable hypertension (*a*) shows a tendency to increase during regular dialysis. No major changes occur in a comparable period in the five patients in whom blood pressure was controlled by regular dialysis (*b*). △, Before regular dialysis; ●, during regular dialysis. From Schalekamp *et al.* (196). (Reproduced by permission of the *American Journal of Medicine*.)

albeit set at a higher level of renin release; the attempt to decrease blood pressure by sodium depletion is thwarted by rising renin.

In the controllable hypertensive state on the other hand, there is evidence that renin is unresponsive to changes in sodium balance (175, 196, 204). It is suggested that this is the reason why blood pressure rises with excess of sodium intake and decreases with sodium deprivation or salt-depleting dialysis (Fig. 19).

There is other evidence for an interaction of renin and body sodium (or plasma volume) in the pathogenesis of renal hypertension. In patients with chronic renal failure the exchangeable sodium is positively correlated with blood pressure as is plasma renin concentration but the product of renin concentration and exchangeable sodium (or plasma volume) correlates better with blood pressure than do either plasma renin or volume alone (199).

In animals, hypertension can be produced by clipping the main renal arteries to one or both kidneys. 'Two-kidney' hypertension is more sodium-dependent than 'one-kidney' hypertension which is more renin-dependent (166, 205). Interestingly, the sodium-dependent form becomes renin-dependent on sodium depletion (205).

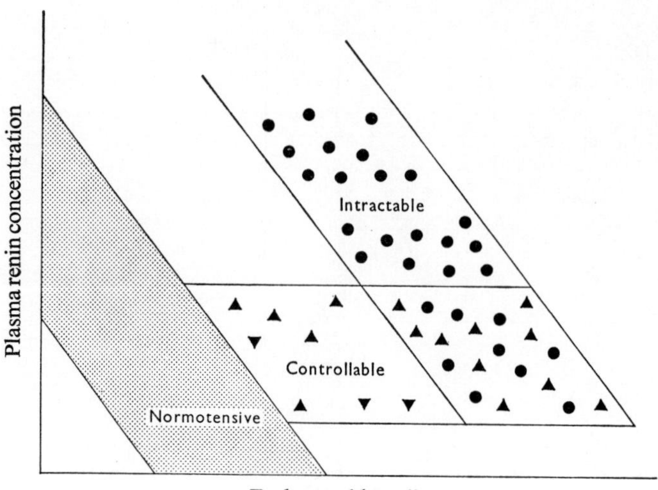

Fig. 19. *Relation between renin and exchangeable sodium in normotensive subjects (as in Fig. 16) compared with the hypothetical relations in controllable and intractable hypertensive states with chronic renal failure (Fig. 18)*

Controllable cases have impaired response of renin to change of sodium balance; in intractable cases the response is abnormally increased. (Reproduced by permission of the *American Journal of Medicine.*)

(*ii*) *Renin concentration normal in relation to exchangeable sodium concentration*

In primary hyperaldosteronism the relation between plasma renin concentration and exchangeable sodium is normal; plasma renin concentration decreases to a value appropriate to the increase of exchangeable sodium (Fig. 20). On lowering of blood pressure by spironolactone or adrenal surgery, sodium is lost, the exchangeable sodium falls and renin concentration rises (116, 206); the relation between exchangeable sodium and renin concentration thus remains normal (Fig. 20).

In patients with essential hypertension and normal renin the relation of renin and exchangeable sodium is also normal.

(*iii*) *Renin concentration low in relation to exchangeable sodium*

In patients with low renin hypertension (see below) or hypertension with deoxycorticosterone excess, plasma renin concentration is significantly low (albeit within the normal range) when related to exchangeable sodium (Fig. 20). Because the values for the exchangeable sodium are close to normal in both groups (Fig. 20), the abnormality would seem to be a suppression of renin secretion which is not associated with increased exchangeable sodium. This contrasts with primary hyperaldosteronism where comparable suppression

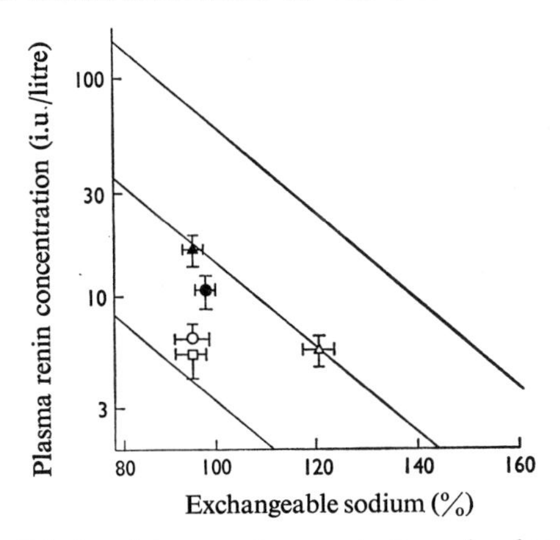

Fig. 20. *Relation of plasma renin concentration and exchangeable sodium*

●, Untreated primary hyperaldosteronism; ▲, treated primary hyperaldosteronism; △, untreated essential hypertension; □, untreated low-renin hypertension; ○, untreated hypertension with deoxycorticosterone excess. Bars represent one S.E.M. for renin (vertical) and sodium (horizontal). Data from Lebel *et al.* (206).

of renin is associated with an expansion of the exchangeable sodium (Figs. 15 and 20).

Low-renin hypertension is a term applied to a group of patients whose plasma renin concentration or renin activity is decreased and fails to respond to stimuli. Between 12 and 45% of patients who would otherwise be classified as essential hypertension share these features (3). Unfortunately, diagnostic criteria are not standardized; different laboratories use different renin assay techniques and different provocative tests (3). Low-renin hypertension has been attributed by several authors to excess of an unidentified sodium-retaining corticosteroid. Evidence favouring this idea is the usually good response of blood pressure to spironolactone (207–209) and the response to aminoglutethamide, an inhibitor of steroid synthesis (210). However, with the provisos about different diagnostic criteria, our own data (Fig. 20 and ref. 206) suggest quite strongly that the suspected sodium retention is not present and this runs counter to the above argument provided the renal site of action of the hypothetical steroid is the same as that of aldosterone. It is possible that excess of a proximally acting steroid could produce hypertension with low plasma renin but that the homeostatic systems (see below) would have decreased the exchangeable sodium more closely to normal than in primary hyperaldosteronism (Fig. 20).

Although the number of patients studied (6) is too small for firm conclusions, the relation of sodium and renin in hypertension with deoxycorticosterone excess is of interest. Again, unlike primary hyperaldosteronism, exchangeable sodium is normal. This suggests either that deoxycorticosterone excess is not responsible for the hypertension or, if it is, that deoxycorticosterone raises blood pressure without generalized sodium retention as manifested by the increased exchangeable sodium. It is of interest that the hypertension produced by injected deoxycorticosterone in rats subjected to unilateral nephrectomy and high sodium intake is associated with increased exchangeable sodium (193).

(*iv*) *Summary*

Blood pressure can be raised separately by excess of body sodium or excess of plasma renin. Conn's syndrome is probably a clinical example of the former and renin-secreting tumour is certainly an example of the latter. However, plasma renin concentration and exchangeable sodium are inversely related in subjects with normal blood pressure and a third possible disorder is an excess of renin in relation to sodium. This could well be important in hypertension with chronic renal failure and in malignant-phase hypertension, but it is clearly not responsible for essential hypertension, low-renin hypertension or hypertension with deoxycorticosterone excess. A possible explanation for the association of a low-renin and a normal exchangeable sodium is discussed on p. 45.

4. Catecholamines and Blood Pressure

There is some evidence that the nervous system and/or circulating catecholamines are involved in essential and renal hypertension (211, 212). Though there is an overlap with the normal range, patients with essential hypertension have been shown, in some studies, to have significantly higher catecholamine levels in urine and blood than normal subjects (32, 213). Also, the plasma concentration of noradrenaline (but not adrenaline) correlates well with blood pressure in essential hypertension and the fall of blood pressure on ganglion blockade is related to the decrease in plasma noradrenaline concentration (214). These observations can be interpreted in several ways. One possibility is that the increase in plasma noradrenaline concentration reflects sympathetic nervous overactivity which is responsible for raising blood pressure.

5. Renin, Catecholamines and the Nervous System

Interesting links have emerged recently between renin and the nervous system; part of the effect of angiotensin on blood vessels may be mediated by the nervous system (215–217). As compared with its effect when given intravenously, angiotensin produces a greater rise of arterial pressure when infused into the cerebral circulation (216, 218, 219). It seems that the sensitive zone lies within the medulla oblongata (220) probably in the area postrema (221). Conversely, the nervous system and/or catecholamines are at least partly involved in renin release (87).

Similar relations could well exist in hypertension (183, 217). In essential hypertension, for example, the change of plasma renin activity on standing is related to the change of urinary catecholamines (213).

6. Renal Abnormality in Essential Hypertension: Cause or Effect?

(*i*) *Maintenance of glomerular filtration by increased filtration fraction*

Excretion of sodium and water by the normal kidney is maintained by the double process of glomerular filtration and tubular reabsorption. Renal blood flow and glomerular filtration generally change in parallel (222). Except in its more advanced stages, the glomerular filtration rate in essential hypertension is well maintained despite a progressive decrease in renal plasma flow (223–226). It follows that filtration fraction (the proportion of plasma filtered) is increased and there is evidence that this increases progressively (225, 226).

(*ii*) *Balance of particular capillary forces*

Recent evidence suggests that small changes in oncotic and hydrostatic pressures within the peritubular capillary circulation have major effects on the reabsorption of sodium and water from the tubule (227–230). If the filtration fraction rises the peritubular oncotic pressure will rise also and sodium reabsorption will be increased (227, 229, 230). Patients with essential hypertension, having an increased filtration fraction, will tend to reabsorb an unusually large proportion of filtered sodium. However, as already described, essential hypertension is not characterized by abnormal sodium retention (Fig. 15). This may be due to the fact that increased blood pressure restores tubular sodium reabsorption and that the patient remains in sodium balance because the blood pressure is raised.

The mechanism of this compensatory effect might also be related to the peritubular circulation. Increased hydrostatic pressure in the peritubular capillaries greatly impairs sodium reabsorption in experimental animals (228, 230). There is evidence from studies of wedged renal venous pressure that peritubular capillary pressure is increased in essential hypertension and that the increase is related to raised arterial pressure transmitted beyond the glomerulus (231). It is possible therefore that raised arterial pressure, by increasing peritubular capillary pressure, restores sodium balance which would otherwise be deranged by the increased filtration fraction.

(*iii*) *Hypothetical mechanism*

It is suggested that over the course of years, renal vascular resistance gradually rises in patients with essential hypertension and that it does so mainly as a result of narrowing in preglomerular vessels (Fig. 21). Glomerular filtration which would otherwise fall is maintained by efferent arteriolar constriction; the filtration fraction increases and this raises the oncotic pressure of peritubular capillary blood; reabsorption of sodium from the renal tubule is

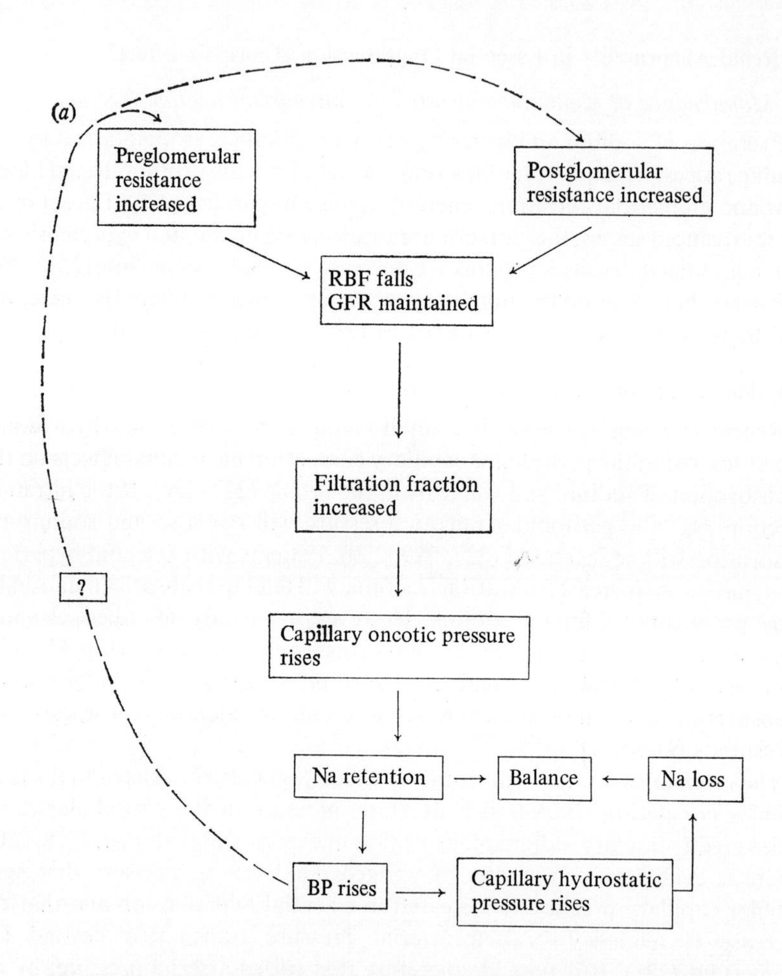

Fig. 21. *Hypothetical mechanism of renal abnormality in essential hypertension*

(*a*) Hypothetical mechanism by which a rise in renal vascular resistance promotes sodium retention which is corrected by a rise of arterial pressure. If the rise of pressure also affects vascular resistance a positive feedback element would be introduced. BP, blood pressure; GFR, glomerular filtration rate; RBF, renal blood flow. (*b*) Hypothetical mechanism in which a rise of arterial pressure leads to increased resistance in the pre- and postglomerular circulation and thence to sodium retention. Sodium balance is restored by the natriuretic effect of increased hydrostatic pressure in the peritubular capillary circulation.

increased and sodium retention leads to increased blood pressure. This is transmitted to the peritubular capillaries where increased hydrostatic pressure decreases sodium reabsorption and restores sodium balance and the exchangeable sodium content to normal albeit at the expense of raised blood pressure.

In this hypothetical mechanism the renal abnormality could be a cause of increased arterial pressure or/and a consequence of increased pressure (cf. Figs. 21*a* and *b*).

(*iv*) *Low plasma renin concentration and progression of the renal abnormality*

It is evident from several studies that plasma renin concentration and renin activity are inversely related to age in essential hypertension (see refs. 3, 225, 226). It may be, therefore, that renin production falls as the patient with essential hypertension ages and that low-renin hypertension is an outcome of this process. The decrease of plasma renin concentration with age is also related to the rise of renal vascular resistance and to the rise of filtration fraction (225, 226). As discussed earlier, renin release may be governed in part by pressure in the afferent arteriole or glomerulus (87). Thus, rising glomerular pressure may be responsible for the depression of renin release, and, if it is, the fall of plasma renin concentration would occur without an increase of total exchangeable sodium or plasma volume (232) (Fig. 20).

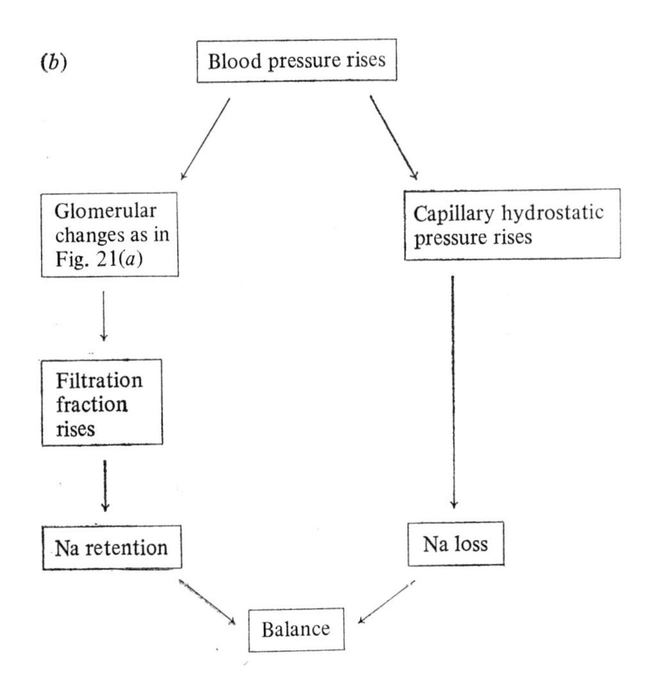

(v) Persistence of hypertension after removal of its cause: a model for essential hypertension

It has been known for many years that removing the causative lesion does not always correct hypertension. Such failures have been noted with most forms of secondary hypertension in man (1, 3) and in animals with renal (162, 233) and deoxycorticosterone (234) hypertension. The duration of hypertension seems important; if blood pressure has been raised for more than a critical period (weeks in the case of renal artery stenosis in the rat) removal of the clipped kidney is ineffective in restoring normotension (162, 233). There is good evidence that exposure of the untouched kidney to hypertension is responsible for this self-perpetuating state of affairs.

Bilateral renal function studies in rats after the application of a unilateral renal artery stenosis and the development of hypertension have shown that the untouched kidney excretes more sodium and water than the clipped kidney (235). Similarities between the 'untouched' kidney of renal hypertension and the kidney of essential hypertension are a suppression of renal renin (236) an enhanced ability to excrete an acute sodium load (235, 237) and an increased filtration fraction (226, 235, 237). Also, the lower the plasma renin concentration in essential hypertension the greater the natriuresis on sodium loading (237). Study of the untouched kidney in renal hypertension may therefore reveal something of the abnormal mechanisms at work in essential hypertension.

If the two processes are identical, the renal functional changes observed in essential hypertension must be a consequence of some other agent which raises blood pressure initially. It has been suspected for some time that so-called 'labile' hypertension is a precursor of 'fixed' essential hypertension (see 3, 9) and this would agree with the mechanism represented in Fig. 21 since glomerular filtration, renal blood flow and filtration fraction are more nearly normal in patients with 'labile' hypertension (225).

(iv) Summary

It is suggested that (*a*) a renal vascular lesion (possibly produced by increased pressure) raises the filtration fraction in essential hypertension, (*b*) sodium retention which would otherwise occur is prevented by a rise of arterial, glomerular and peritubular capillary pressure (*c*) the decrease of plasma renin is due to the increased glomerular pressure.

It is also suggested that the untouched kidney in unilateral renal artery stenosis may be a model for these events and that the ability of this kidney to maintain hypertension has a similar mechanism.

Conclusions and General Summary

Though impossible to define in other than arbitrary terms, hypertension is common, dangerous and treatable. Because it is usually symptomless, there may be a case for population screening, but the real benefits of this have yet to be assessed.

Distinguishing the various syndromes with hypertension is relatively simple and the contribution of the biochemist to this is important. Attempts to define the mechanism by which blood pressure rises in these syndromes have been less successful with the exception of renin-secreting tumour, phaeochromocytoma and primary hyperaldosteronism.

An abnormal relation between two mechanisms, sodium and the renin–angiotensin system, may raise the blood pressure in renal hypertension. Essential hypertension remains an enigma, though an abnormality of renal function leading to a resetting of the pressure–natriuretic mechanism may be important. There could be an interesting analogy between this and the 'untouched' kidney in hypertension due to unilateral renal artery stenosis.

Note Added in Proof

Semple *et al.* (249, 250) have recently developed a method for extracting and measuring separately angiotensin II, its heptapeptide fragment (so-called angiotensin III) and the hexapeptide/pentapeptide fragments. Contrary to earlier views (252), hepta- and hexa-peptide fragments are not present in important amounts in human venous plasma. On the other hand, the heptapeptide is present in greater amount than the octapeptide in rat plasma (251). This is of interest, as the heptapeptide is capable of stimulating aldosterone secretion in the rat (253), and the role of angiotensin II as a stimulus of aldosterone in the rat has been in doubt for some time. Angiotensin III may be more important.

Technical aspects of the assay of renin, renin activity, angiotensin I and angiotensin II have been reviewed recently (78, 84). It is proposed shortly to introduce an international renin standard for use in assay of renin concentration and renin activity (254). One international unit is equivalent to 190 units in our own assay.

References

(1) Pickering, G. W. (1968) *High Blood Pressure*, 2nd edn., J. and A. Churchill, London
(2) Pickering, G. W. (1970) *Hypertension: Causes, Consequences and Management*, J. and A. Churchill, London
(3) Kaplan, N. M. (1973) *Clinical Hypertension*, Medcom Press, New York
(4) Genest, J. & Koiw, E. (1972) *Hypertension 1972*, Springer-Verlag, Berlin
(5) Robertson, J. I. S. (ed.) (1973) *Clin. Sci. Mol. Med.* **45**, *Suppl.* **1**: *Proc. 3rd Meet. Int. Soc. Hypertension*
(6) Laragh, J. H. (ed.) (1972) *Amer. J. Med.* **52**, 565–678
(7) Laragh, J. H. (ed.) (1973) *Amer. J. Med.* **53**, 261–402
(8) *Circulation Res.* Annual supplements are issued on hypertension
(9) Brown, J. J., Fraser, R., Lever, A. F. & Robertson, J. I. S. (1971) Hypertension: a review of selected topics. *Abstr. World Med.* **45**, 549–559, 633–644

(10) Grimley Evans, J. & Rose, G. (1971) Hypertension. *Brit. Med. Bull.* **27**, 37–42
(11) Dollery, C. T. (1973) Adrenergic drugs in the treatment of hypertension. *Brit. Med. Bull.* **29**, 158–162
(12) Prichard, B. N. C. (1973) Hypotensive agents. *Brit. J. Hosp. Med.* July issue, 45–52
(13) Freis, E. D. (1973) Age, race, sex and other indices of risk in hypertension. *Amer. J. Med.* **55**, 275–280
(14) Rose, G. (1973) The treatment of mild hypertension. *Symp. Advan. Med. 9th* (Walker, G., ed.), pp. 219–226, Pitman Medical, London
(15) Hawthorne, V. M., Greaves, D. A. & Beevers, D. G. (1974) Blood pressure in a Scottish town. *Brit. Med. J.* **3**, 600–603
(16) Peart, W. S. (1973) Organisation of a multicentre randomized-control therapeutic trial for mild to moderate hypertension. *Clin. Sci. Mol. Med.* **45**, 67s–70s
(17) Molinoff, P. B. & Axelrod, J. (1971) Biochemistry of catecholamines. *Annu. Rev. Biochem.* **40**, 465–500
(18) Blascho, H. (1972) Catecholamine biosynthesis. *Brit. Med. Bull.* **29**, 105–109
(19) Harrison, T. S., Chawla, R. C. & Wojtalik, R. S. (1968) Steroidal influences on catecholamines. *N. Engl. J. Med.* **279**, 136–143
(20) Dobbie, J. W. & Symington, T. (1966) The human adrenal gland with special reference to the vasculature. *J. Endocrinol.* **34**, 479–489
(21) Jenkinson, D. H. (1972) Classification and properties of peripheral adrenergic receptors. *Brit. Med. Bull.* **29**, 142–147
(22) Allwood, M. J., Cobbold, A. F. & Ginsburg, J. (1963) Peripheral vascular effects of noradrenaline, isopropyl noradrenaline and dopamine. *Brit. Med. Bull.* **19**, 132–136
(23) Ross, E. J. (1973) Diseases of the adrenal medulla: phaeochromocytoma. *Medicine* **2**, 157–160
(24) Symington, T. (1969) Functional pathology of the human adrenal gland. Livingstone, London
(25) Brown, J. J. (1972) Phaeochromocytoma. In *Scientific Basis of Surgery*, 2nd edn. (Irvine, W. T., ed.), pp. 311–332, Churchill Livingstone, London
(26) Hume, D. M. (1960) Phaeochromocytoma in the adult and in the child. *Amer. J. Surg.* **99**, 458–496
(27) Robinson, R., Smith, P. & Whittaker, S. R. F. (1964) Secretion of catecholamines in malignant phaeochromocytoma. *Brit. Med. J.* **1**, 1422–1424
(28) Williams, E. D., Brown, C. L. & Doniach, I. (1966) Pathological and clinical findings in a series of 67 cases of medullary carcinoma of thyroid. *J. Clin. Pathol.* **19**, 103–113
(29) Williams, E. D. & Pollock, D. J. (1966) Multiple mucosal neuromata with endocrine tumour syndrome allied to Von Recklinghausen's disease. *J. Pathol. Bacteriol.* **9**, 71–80
(30) Editorial (1965) Pressor attacks during treatment with monoamine-oxidase inhibitors. *Lancet* **i**, 945–946
(31) Hunyor, S. N., Hansson, L., Harrison, T. S. & Hoobler, S. W. (1973) Effects of clonidine withdrawal: possible mechanisms and suggestions for management. *Brit. Med. J.* **2**, 209–211
(32) Engelman, K., Portnoy, B. & Sjoerdsma, A. (1970) Plasma catecholamine concentration in patients with hypertension. *Circ. Res.* **27**, *Suppl.* **1**, 141–146
(33) Geffen, L. B., Rush, R. A., Louis, W. J. & Doyle, A. E. (1973) Plasma catecholamine and dopamine β-hydroxylase amounts in phaeochromocytoma. *Clin. Sci.* **44**, 421–424
(34) Callingham, B. A. (1967) The catecholamines and adrenaline; noradrenaline. In *Hormones in Blood* (Gray, C. H. & Bacharach, A. L., eds.), vol. 2, pp. 519–599, Academic Press, London
(35) Carruthers, M., Conway, N., Taggart, P., Bates, D. & Sommerville, W. (1970) Validity of plasma catecholamine estimations. *Lancet* **ii**, 62–67
(36) Passon, P. G. & Peuler, J. D. (1973) A simplified radiometric assay for plasma norepinephrine and epinephrine. *Anal. Biochem.* **51**, 618–631

(37) Wong, K. P., Ruthven, C. R. J. & Sandler, M. (1973) Gas chromatographic measurement of urinary catechol amines by an electron capture detection procedure. *Clin. Chim. Acta* **47**, 215–222
(38) Sheps, S. G., Tyce, G. M., Flock, E. V. & Maher, F. T. (1966) Current experience in the diagnosis of phaeochromocytoma. *Circulation* **34**, 473–483
(39) Porter, G. A. & Kimsey, J. (1971) Assessment of the mineralocorticoid activity of 18-hydroxy-11-deoxycorticosterone in the isolated toad bladder. *Endocrinology* **89**, 353–357
(40) Samuels, L. T. & Uchikawa, T. (1967) *Biosynthesis of Adrenal Steroids in the Adrenal Cortex* (Eisenstein, A. B., ed.), pp. 61–102, Churchill, London
(41) Pasqualini, J. R. (1964) Conversion of tritiated 18-hydroxycorticosterone to aldosterone by slices of human cortico-adrenal galnd and adrenal tumour. *Nature* (*London*) **201**, 501
(42) Vecsei, P. & Glaz, E. (1971) *Aldosterone* Pergamon, Oxford
(43) James, V. H. T. & Landon, J. (1968) The control of corticosteroid secretion: current views and methods of assessment. In *Recent Advances in Endocrinology* (James, V. H. T., ed.) pp. 50–94, Churchill, London
(44) Blackwell, R. E. & Guillemin, R. (1973) Hypothalamic control of adenohypophyseal secretions. *Annu. Rev. Physiol.* **35**, 357–390
(45) Stone, D. & Hechter, O. (1954) Studies on ACTH action in perfused bovine adrenal glands. Site of action of ACTH in corticosteroidogenesis. *Arch. Biochim. Biophys.* **51**, 457–469
(46) Yates, F. E. & Urquhart, J. (1962) Control of plasma concentrations of adrenocortical hormones. *Physiol. Rev.* **42**, 359–443
(47) Sutherland, E. W. & Robison, G. A. (1966) The role of cyclic AMP in response to catecholamines and other hormones. *Pharmacol. Rev.* **18**, 145–161
(48) Major, P. W. & Kilpatrick, R. (1972) Cyclic AMP and hormone action. *J. Endocrinol.* **52**, 593–630
(49) Garren, L. D., Gill, G. N., Masui, H. & Walton, G. M. (1971) On the mechanism of action of ACTH. *Recent Progr. Horm. Res.* **27**, 433–478
(50) Li Choh Hao & Oelofsen, W. (1967) The chemistry and biology of ACTH and related peptides. In *The Adrenal Cortex* (Eisenstein, A. B., ed.), pp. 185–201, Churchill, London
(51) Landon, J. (1968) The basis of radioimmunoassay with particular reference to adrenocorticotrophic hormone. In *Recent Advances in Endocrinology* (James, V. H. T., ed.), pp. 240–270, Churchill, London
(52) Besser, G. M. & Edwards, C. R. W. (1972) Cushing's syndrome. *Clin. Endocrinol. Metab.* **1**, 451–490
(53) Chayen, J., Loveridge, N. & Daly, J. R. (1972) A sensitive bioassay for adrenocorticotrophic hormone in human plasma. *Clin. Endocrinol.* **1**, 219–233
(54) Alaghband-Zadeh, J., Daly, J. R., Tunbridge, R. D. G., Loveridge, N. & Chayen, J. (1973) Methodological refinements in the redox assay for ACTH. *J. Endocrinol.* **58**, xix
(55) Ross, E. J., Marshak-Jones, P. & Friedman, M. (1966) Cushing's syndrome: diagnostic criteria. *Quart. J. Med.* **35**, 149–192
(56) Plotz, C. M., Knowlton, A. I. & Ragan, C. (1952) The natural history of Cushing's syndrome. *Amer. J. Med.* **13**, 597–614
(57) Liddle, G. W. (1967) Cushing's syndrome. In *The Adrenal Cortex* (Eisenstein, A. B., ed.), pp. 523–551. Churchill, London
(58) James, V. H. T., Landon, J., Wynn, V. & Greenwood, F. C. (1968) A fundamental defect of adrenocortical control in Cushing's disease. *J. Endocrinol.* **40**, 15–28
(59) Ratcliffe, J. G., Knight, R. A., Besser, G. M., Landon, J. & Stansfeld, A. G. (1973) Tumour and plasma ACTH concentrations in patients with and without ectopic ACTH syndrome. *Clin. Endocrinol.* **1**, 27–44
(60) Burke, C. W., Doyle, F. H., Joplin, G. F., Arnot, R. N., Macerlean, D. P. & Russell Fraser, T. (1973) Cushing's disease. Treatment by pituitary implantation of radioactive gold or yttrium seeds. *Quart. J. Med.* **42**, 693–714

(61) Ross, E. J. (1968) The cancer cell as an endocrine organ. In *Recent Advances in Endocrinology*, 8th edn. (James, V. H. T., ed.), pp. 293–327, Churchill, London
(62) Anderson, G. (1973) Paramalignant syndromes. In *Recent Advances in Medicine* (Baron, D. N., Compston, N. & Dawson, A. M., eds.), pp. 1–29, Churchill Livingstone, Edinburgh
(63) Monro, D. D. & Clift, D. C. (1973) Pituitary–adrenal function after prolonged use of topical corticosteroids. *Brit. J. Dermatol.* **88**, 381–385
(64) Burke, C. W. & Beardwell, C. G. (1973) Cushing's syndrome. An evaluation of the clinical usefulness of urinary free cortisol and other urinary steroid measurements in diagnosis. *Quart. J. Med.* **42**, 175–204
(65) Eddy, R. L., Lloyd-Jones, A., Gilliland, P. F., Ibarra, J. D., Thompson, J. Q. & McMurry, J. F. (1973) Cushing's syndrome: a prospective study of diagnostic methods. *Amer. J. Med.* **55**, 621–630
(66) Cope, C. L. (1964) *Adrenal Steroids and Disease* Pitman, London
(67) Butler, P. W. P. & Besser, G. M. (1968) Pituitary–adrenal function in severe depressive illness. *Lancet* **i**, 1234–1236
(68) Scoggins, B. A., Coghlan, J. P., Cran, E. J., Denton, D. A., Fan, S. K., McDougall, J. G., Oddie, C., Robinson, D. M. & Shulkes, A. A. (1973) Experimental studies on the mechanism of adrenocorticotrophic hormone-induced hypertension in the sheep. *Clin. Sci. Mol. Med.* **45**, 269 s–271 s
(69) Biglieri, E. G., Slaton, P. E., Schambelan, M. & Kronfield, S. J. (1968) Hypermineralocorticoidism. *Amer. J. Med.* **45**, 170–175
(70) Peart, W. S. (1965) The renin–angiotensin system. *Pharm. Rev.* **17**, 143–182
(71) Brown, J. J., Davies, D. L., Lever, A. F. & Robertson, J. I. S. (1966) Renin and angiotensin: A survey of some aspects. *Postgrad. Med. J.* **42**, 153–176
(72) Cook, W. F. (1971) in *Kidney Hormones* (Fisher, J. W., ed.), pp. 117–128, Academic Press, London and New York
(73) Leckie, B. (1973) The activation of a possible zymogen of renin in rabbit kidney. *Clin. Sci.* **44**, 301–304
(74) Skeggs, L. T., Lentz, K. E., Hochstrasser, H. & Kahn, J. R. (1964) The chemistry of renin substrate. *Can. Med. Ass. J.* **90**, 185–189
(75) Vane, J. R. (1969) The release and fate of vaso-active hormones in the circulation. *Brit. J. Pharmacol.* **45**, 209–242
(76) Oparil, S., Treagear, G. W., Koerner, T., Barnes, B. A. & Haber, E. (1971) Mechanism of pulmonary conversion of angiotensin I to angiotensin II in the dog. *Circ. Res.* **29**, 682–690
(77) Volker, A., Kreye, W. & Gross, F. (1971) Conversion of angiotensin I to angiotensin II in peripheral vascular beds of the rat. *Amer. J. Physiol.* **220**, 1294–1296
(78) Morton, J. J., Waite, M. A., Brown, J. J., Lever, A. F., Robertson, J. I. S. & Semple, P. F. (1975) The estimation of angiotensin I and II in the human circulation by radioimmunoassay. In *Hormones in Human Plasma* (Antoniades, H. N., ed.), Academic Press, London and New York, in the press
(79) Skeggs, L. T., Kahn, J. R. & Shumway, N. P. (1956) Preparation and function of the hypertensin converting enzyme. *J. Exp. Med.* **103**, 295–299
(80) Fitzsimons, J. T. (1972) Thirst. *Physiol. Rev.* **52**, 468–561
(81) Chinn, R. H. & Düsterdieck, G. (1971) The response of blood pressure to infusion of angiotensin II; relation to plasma concentrations of renin and angiotensin II. *Clin. Sci.* **42**, 489–504
(82) Oelkers, W., Brown, J. J., Fraser, R., Lever, A. F., Morton, J. J. & Robertson, J. I. S. (1974) Sensitization of the adrenal cortex to angiotensin II in sodium-deplete man. *Circ. Res.* **34**, 69–77
(83) Macgregor, J., Briggs, J. D., Brown, J. J., Chinn, R. H., Gavras, H., Lever, A. F., Macadam, R. F., Medina, A., Morton, J. J., Oliver, N. W. J., Paton, A., Powell-Jackson, J. D., Robertson, J. I. S. & Waite, M. A. (1973) Renin and renal function. In *Modern Diuretic Therapy in the Treatment of Cardiovascular and Renal Disease* (Lant, A. F. & Wilson, G. M., eds.), pp. 71–82, Excerpta Medica, Amsterdam

(84) Poulsen, K. (1973) Renin and aldosterone. Measurements of renin–angiotensin–aldosterone. *Pharm. Rev.* **25**, 249–259

(85) Blair-West, J. R., Coghlan, J. P., Denton, D. A., Funder, J. W., Scoggins, B. A. & Wright, R. D. (1971) The effect of the heptapeptide (2–8) and hexapeptide (3–8) fragments of angiotensin II on aldosterone secretion. *J. Clin. Endocrinol.* **32**, 575–578

(86) Hosie, K. F., Brown, J. J., Harper, A. M., Lever, A. F., Macadam, R. F., MacGregor, J. & Robertson, J. I. S. (1970) The release of renin into the renal circulation of the anaesthetized dog. *Clin. Sci.* **38**, 157–174

(87) Davis, J. O. (1973) The control of renin release. *Amer. J. Med.* **55**, 333–350

(88) Thurau, K. C. W., Dahlheim, H., Gruner, A., Mason, J. & Granger, P. (1972) Activation of renin in the single juxtaglomerular apparatus by sodium chloride in the tubular fluid at the macula densa. *Circ. Res.* **31**, *Suppl.* **2**, 182–186

(89) Brown, J. J., Fraser, R., Lever, A. F. & Robertson, J. I. S. (1968) in *Recent Advances in Endocrinology*, 8th edn. (James, V. H. T., ed.), pp. 271–292, Churchill, London

(90) Brown, J. J., Lever, A. F., Morton, J. J., Fraser, R., Love, D. R. & Robertson, J. I. S. (1972) Raised plasma angiotensin II and aldosterone during dietary sodium restriction in man. *Lancet* **II**, 1106–1107

(91) Kassirer, J. P., London, A. M., Goldman, D. M. & Schwartz, W. B. (1970) On the pathogenesis of metabolic alkalosis in hyperaldosteronism. *Amer. J. Med.* **49**, 306–313

(92) Edelman, I. & Fimognari, G. M. (1968) Biochemical mechanisms of the action of aldosterone. *Recent Progr. Horm. Res.* **24**, 1–44

(93) Fraser, R. (1971) The effect of steroids on the transport of electrolytes through membranes. *Biochem. Soc. Symp.* **32**, 101–127

(94) Birkhauser, M., Gaillard, R., Riondel, A. M., Scholer, D., Vallotton, M. B. & Muller, A. F. (1973) Effect of volume expansion by hyperosmolar and hyperoncotic solutions under constant infusion of angiotensin II on plasma aldosterone in man and its counterbalance by potassium administration. *Eur. J. Clin. Invest.* **3**, 307–316

(95) Baumber, J. S., Davis, J. O., Johnson, J. A. & Witty, R. T. (1971) Increased adrenocortical potassium in association with increased biosynthesis of aldosterone. *Amer. J. Physiol.* **220**, 1094–1099

(96) Blair-West, J. R., Cain, M. D., Catt, K. J., Coghlan, J. P., Denton, D. A., Funder, J. W., Scoggins, B. A., Wintour, M. & Wright, R. D. (1970) The mode of control of aldosterone secretion. *Proc. Int. Congr. Nephrol. 4th* **2**, 33–44

(97) Dluhy, R. G., Lloyd, A., Underwood, R. H. & Williams, G. H. (1972) Studies of the control of plasma aldosterone concentration in normal man. II. Effect of dietary potassium and acute potassium infusion. *J. Clin. Invest.* **51**, 1950–1957

(98) Scholer, D., Birkhauser, M., Peytremann, A., Riondel, A. M., Vallotton, M. B. & Muller, A. F. (1973) Response of plasma aldosterone to angiotensin II, ACTH and potassium in man. *Acta Endocrinol.* (*Copenhagen*) **72**, 293–307

(99) Williams, G. H., Dluhy, R. G. & Underwood, R. H. (1970) The relationship of dietary potassium intake to the aldosterone stimulating properties of ACTH. *Clin. Sci.* **39**, 489–496

(100) Brown, J. J., Fraser, R., Lever, A. F. & Robertson, J. I. S. (1972) Aldosterone: physiological and pathophysiological variations in man. *Clin. Endocrinol. Metab.* **1**, 397–449

(101) Blair-West, J. R., Coghlan, J. P., Cran, E., Denton, D. A., Funder, J. W. & Scoggins, B. A. (1972) Contrived suppression of renin secretion during sodium depletion. In *Hypertension* 1972 (Genest, J. & Koiw, E., eds.), pp. 14–25, Springer-Verlag, Berlin

(102) Boyd, G. W., Adamson, A. R., Arnold, M., James, V. H. T. & Peart, W. S. (1972) The role of angiotensin II in the control of aldosterone in man. *Clin. Sci.* **42**, 91–104

(103) Gross, F. (1958) Renin und Hypertensin, Physiologische oder Pathologische Wirkstoffe. *Klin. Wochenschr.* **36**, 693–706

(104) Brown, J. J., Davies, D. L., Lever, A. F., Robertson, J. I. S. & Tree, M. (1964) The estimation of renin in human plasma. *Biochem. J.* **93**, 594–600

(105) Stockigt, J. R., Collins, R. D. & Biglieri, E. G. (1971) Determination of plasma renin concentration by angiotensin I radioimmunoassay. *Circ. Res.* **28**, *Suppl.* **2**, 175–189

(106) Sealey, J. E. & Laragh, J. L. (1973) Searching out low renin patients: limitations of some commonly used methods. *Amer. J. Med.* **55**, 303–313

(107) Tree, M. (1973) Measurement of plasma renin-substrate in man. *J. Endocrinol.* **56**, 159–171

(108) Waite, M. A., Tree, M. & McDermott, E. (1973) The estimation of human renin-substrate concentration by radioimmunoassay of angiotensin I. *J. Endocrinol.* **57**, 329–330

(109) Waite, M. A. (1973) Measurement of concentrations of angiotensin I in human blood by radioimmunoassay. *Clin. Sci.* **45**, 51–64

(110) Fraser, R., Guest, S. & Young, J. (1973) A comparison of double-isotope derivative and radioimmunological estimation of plasma aldosterone concentration in man. *Clin. Sci. Mol. Med.* **45**, 411–415

(111) Coghlan, J. P. & Blair-West, J. R. (1967) Aldosterone. In *Hormones in Blood* (Gray, C. H. & Bacharach, A. L., eds.), vol. 2, pp. 391–488, Academic Press, London

(112) Oddie, C. J., Coghlan, J. P. & Scoggins, B. A. (1972) Plasma deoxycorticosterone levels in man with simultaneous measurement of aldosterone, corticosterone, cortisol and 11-deoxycortisol. *J. Clin. Endocrinol. Metab.* **34**, 1039–1054

(113) Skinner, S. L., Lumbers, E. R. & Symonds, E. M. (1972) Analysis of changes in the renin–angiotensin system during pregnancy. *Clin. Sci.* **42**, 479–488

(114) Robertson, J. I. S., Weir, R. J., Dusterdieck, G. O., Fraser, R. & Tree, M. (1971) Renin, angiotensin and aldosterone in human pregnancy and the menstrual cycle. *Scot. Med. J.* **16**, 183–196

(115) Brown, J. J., Fraser, R., Lever, A. F. & Robertson, J. I. S. (1972) Hypertension with aldosterone excess. *Brit. Med. J.* **2**, 391–396

(116) Beevers, D. G., Brown, J. J., Ferriss, J. B., Fraser, R., Lever, A. F. & Robertson, J. I. S. (1973) The use of spironolactone in the diagnosis and the treatment of hypertension associated with mineralocorticoid excess. *Amer. Heart J.* **86**, 404–414

(117) Edmonds, C. J. & Jasani, B. (1972) Total body potassium in hypertensive patients during prolonged diuretic therapy. *Lancet* **ii**, 8–12

(118) Tarazi, R. C., Dustan, H. P. & Frohlich, E. D. (1970) Long-term thiazide therapy in essential hypertension. *Circulation* **41**, 709–717

(119) Conn, J. W. (1955) Primary aldosteronism. *J. Lab. Clin. Med.* **45**, 661–664

(120) Brown, J. J., Chinn, R. H., Davies, D. L., Düsterdieck, G., Fraser, R., Lever, A. F., Robertson, J. I. S., Tree, M. & Wiseman, A. (1968) Plasma electrolytes, renin and aldosterone in the diagnosis of primary hyperaldosteronism. *Lancet* **ii**, 55–59

(121) Conn, J. W., Cohen, E. L., Rovner, D. R. & Nesbit, R. M. (1965) Normokalaemic primary aldosteronism. *J. Amer. Med. Ass.* **193**, 200–206

(122) Biglieri, E. D. & Forsham, P. H. (1961) Studies on the expanded extracellular fluid and the responses to various stimuli in primary aldosteronism. *Amer. J. Med.* **30**, 564–576

(123) Brown, J. J., Davies, D. L., Lever, A. F., Peart, W. S. & Robertson, J. I. S. (1965) Plasma concentration of renin in a patient with Conn's syndrome with fibrinoid lesions of the renal arterioles: the effect of treatment with spironolactone. *J. Endocrinol.* **33**, 279–293

(124) Biglieri, E. G., Stockigt, J. R. & Schambelan, M. (1972) Adrenal mineralocorticoids causing hypertension. *Amer. J. Med.* **52**, 623–632

(125) Aitchison, J., Brown, J. J., Ferriss, J. B., Fraser, R., Kay, A. W., Lever, A. F., Neville, A. M., Symington, T. & Robertson, J. I. S. (1971) Quadric analysis in the preoperative distinction between patients with and without adrenocortical tumours in hypertension with aldosterone excess and low plasma renin. *Amer. Heart. J.* **82**, 660–671

(126) Spark, R. F., & Melby, J. C. (1968) Aldosteronism in hypertension. The spironolactone response test. *Ann. Intern. Med.* **69**, 685–691

(127) Kremer, D., Beevers, D. G., Brown, J. J., Davies, D. L., Ferriss, J. B., Fraser, R., Lever, A. F. & Robertson, J. I. S. (1973) Spironolactone and amiloride in the treatment of low renin hyperaldosteronism and related syndromes. *Clin. Sci. Mol. Med.* **45**, 213s–218s

(128) Ferriss, J. B., Brown, J. J., Fraser, R., Haywood, E., Davies, D. L., Kay, A. W., Lever, A. F., Robertson, J. I. S., Owen, K. & Peart, W. S. (1975) Results of adrenal surgery in patients with hypertension, aldosterone excess and low plasma renin concentration. *Brit. Med. J.* **1**, 135–141

(129) Kem, D. C., Weinberger, M. H., Gomez-Sanchez, C., Kramer, N. J., Lerman, R., Furuyama, S. & Nugent, C. A. (1973) Circadian rhythm of plasma aldosterone concentration in patients with primary aldosteronism. *J. Clin. Invest.* **52**, 2272–2277

(130) Ganguly, A., Dowdy, A. J., Luetscher, J. A. & Melada, G. A. (1973) Anomalous postural response of plasma aldosterone concentration in patients with aldosterone-producing adrenal adenoma. *J. Clin. Endocrinol. Metab.* **36**, 401–404

(131) Brooks, R. U., Jeffcoate, S. L., London, D. R., Prunty, F. T. G. & Smith, D. M. (1966) Intermittent Cushing's syndrome with anomalous response to dexamethasone. *J. Endocrinol.* **36**, 53–61

(132) Gowenlock, A. H. & Wrong, O. (1962) Hyperaldosteronism secondary to renal ischaemia. *Quart. J. Med.* **31**, 323–343

(133) Bloch, H. S. (1966) Hypertension, bilateral renal artery stenosis, adrenocortical adenomas and normal serum electrolyte levels. *J. Amer. Med. Ass.* **196**, 622–624

(134) Brown, J. J., Davies, D. L., Lever, A. F. & Robertson, J. I. S. (1963) Plasma renin in relation to alterations in sodium metabolism. In *Boerhaave Course: Hypertension* (de Graeff, J., ed.), pp. 216–221, Leiden

(135) Giebink, G. S., Gotlin, R. W., Biglieri, E. G. & Katz, F. H. (1973) A kindred with familial glucocorticoid-suppressible aldosteronism. *J. Clin. Endocrinol. Metab.* **36**, 715–743

(136) Sutherland, D. J. A., Ruse, J. L. & Laidlaw, J. C. (1966) Hypertension, increased aldosterone secretion, and low plasma renin activity relieved by dexamethasone. *Can. Med. Ass. J.* **95**, 1109–1119

(137) New, M. I., Siegal, E. J. & Peterson, R. E. (1973) Dexamethasone-suppressible hyperaldosteronism. *J. Clin. Endocrinol. Metab.* **37**, 93–100

(138) Barraclough, M. A., Bacchus, B., Brown, J. J., Davies, D. L., Lever, A. F. & Robertson, J. I. S. (1965) Plasma renin and aldosterone secretion in hypertensive patients with renal or renal artery lesions. *Lancet* **ii**, 1310–1313

(139) Brown, J. J., Düsterdieck, G., Fraser, R., Lever, A. F., Robertson, J. I. S., Tree, M. & Weir, R. J. (1971) Hypertension and chronic renal faiure. *Brit. Med. Bull.* **27**, 128–135

(140) Editorial (1973) Primary excess and deficiency of renin. *Brit. Med. J.* **1**, 627–628

(141) Robertson, J. I. S., Brown, J. J., Düsterdieck, G. O., Ferriss, J. B., Fraser, R. & Lever, A. F. (1972) Hypertension with aldosterone excess. *Anglo-Ger. Med. Rev.* **6**, 55–73

(142) Editorial (1973) Problems in renovascular hypertension. *Brit. Med. J.* **4**, 566–568

(143) Hall, C. E. & Hall, O. (1965) Hypertension and hypersalimentation. II. Deoxycorticosterone hypertension. *Lab. Invest.* **14**, 1727–1735

(144) Beilin, L. J., Wade, D. N., Honour, A. J. & Cole, T. J. (1970) Vascular hyper-reactivity with sodium loading and with desoxycorticosterone induced hypertension in the rat. *Clin. Sci.* **39**, 793–810

(145) Wilson, A. & Fraser, R. (1971) The estimation of plasma 11-deoxycorticosterone in man using gas–liquid chromatography with electron capture detection. *J. Endocrinol.* **51**, 557–567

(146) Brown, J. J., Ferriss, J. B., Fraser, R., Lever, A. F., Love, D. R., Robertson, J. I. S. & Wilson, A. (1972) Apparently isolated excess deoxycorticosterone in hypertension, another variant of the mineralocorticoid excess syndrome. *Lancet* **ii**, 243–247

(147) Melby, J. C., Dale, S. L. & Wilson, T. E. (1971) 18-Hydroxydeoxycorticosterone in human hypertension. *Circ. Res.* **28**, *Suppl.* **2**, 143–150

(148) Genest, J., Nowaczynski, W., Kuchel, O. & Sasaki, C. (1972) Plasma progesterone levels and 18-hydroxydeoxycorticosterone secretion rate in benign essential hypertension in humans. In *Hypertension* 1972 (Genest, J. & Koiw, E., eds.), pp. 293–298, Springer Verlag, Berlin

(149) Fraser, R., James, V. H. T., Landon, J., Peart, W. S., Rawson, A., Giles, C. A. & McKay, A. M. (1968) Clinical and biochemical studies of a patient with a corticosterone-secreting adrenocortical tumour. *Lancet* **ii**, 1116–1120

(150) Biglieri, E. G., Herron, M. A. & Brust, N. (1966) 17-Hydroxylation deficiencies in man. *J. Clin. Invest.* **45**, 1946–1954

(151) Bricaire, H., Luton, J. P., Laudat, P., Legrand, J. C., Turpin, G., Coruol, P. & Lemmer, M. (1972) A new male pseudohermaphroditism associated with hypertension due to a block of 17-α-hydroxylation. *J. Clin. Endocrinol. Metab.* **35**, 67–72

(152) Goldsmith, O., Solomon, D. H. & Horton, R. (1967) Hypogonadism and mineralocorticoid excess: the 17-hydroxylase deficiency syndrome. *N. Engl. J. Med.* **277**, 673–677

(153) New, M. I. & Seaman, M. P. (1970) Secretion rates of cortisol and aldosterone precursors in various forms of congenital adrenal hyperplasia. *J. Clin. Endocrinol. Metab.* **30**, 361–371

(154) Borst, J. G. G. & Borst-de Geus, A. (1963) Hypertension explained by Starling's theory of circulatory homeostasis. *Lancet* **i**, 677–682

(155) Conn, J. W., Rovner, D. R. & Cohen, E. L. (1968) Licorice-induced pseudoaldosteronism. *J. Amer. Med. Ass.* **205**, 492–496

(156) Mohamed, S. D., Chapman, R. S. & Crooks, J. (1966) Hypokalaemia, flaccid quadruparesis and myoglobinuria with carbenoxolone (Biogastrone). *Brit. Med. J.* **1**, 1581–1582

(157) Mayes, D., Furuyama, S., Kem, D. C. & Nugent, C. A. (1970) A radioimmunoassay for plasma aldosterone. *J. Clin. Endocrinol.* **30**, 682–685

(158) Baron, J. H. & Nabarro, J. D. N. (1968) Metabolic studies of carbenoxolone sodium as biogastrone and duogastrone in patients with peptic ulcer. In *A Symposium on Carbenoxolone Sodium* (Robson, J. M. & Sullivan, F. M., eds.), pp. 127–157, Butterworths, London

(159) Wilson, C. & Byrom, F. B. (1939) Renal changes in malignant hypertension. *Lancet* **i**, 136–139

(160) Byrom, F. B. & Dodson, L. F. (1949) Mechanism of the vicious circle in chronic hypertension. *Clin. Sci.* **8**, 1–10

(161) Floyer, M. A. (1951) The effect of nephrectomy and adrenalectomy upon the blood pressure in hypertensive and normotensive rats. *Clin. Sci.* **10**, 405–421

(162) Koletsky, S. & Riuera-Velez, J. M. (1970) Factors determining the success or failure of nephrectomy in experimental renal hypertension. *J. Lab. Clin. Med.* **76**, 54–65

(163) Thal, A. P., Grage, T. B. & Vernier, R. L. (1963) Function of the contralateral kidney in renal hypertension due to renal artery stenosis. *Circulation* **27**, 36–43

(164) Hunt, J. C. & Strong, C. G. (1973) Renovascular hypertension, mechanisms, natural history and treatment. *Amer. J. Cardiol.* **32**, 562–574

(165) Capelli, J. P., Wesson, L. G. & Housel, E. L. (1973) Renovascular hypertension: incidence, diagnosis, mechanism and treatment. *J. Chronic. Dis.* **26**, 503–527

(166) Bianchi, G., Baldoni, E., Lucca, R. & Barbin, P. (1972) Pathogenesis of arterial hypertension after the construction of the renal artery leaving the opposite kidney intact both in the anaesthetised and in the conscious dog. *Clin. Sci.* **42**, 651–664

(167) Brown, J. J., Peart, W. S., Owen, K., Robertson, J. I. S. & Sutton, D. (1960) The diagnosis and treatment of renal artery stenosis. *Brit. Med. J.* **2**, 327–338

(168) Barraclough, M. A. (1966) Sodium and water depletion with acute malignant hypertension. *Amer. J. Med.* **40**, 265–272

(169) Vaughan, E. D., Bühler, F. R., Laragh, J. H., Sealey, J. E., Baer, L. & Bard, R. H. (1973) Renovascular hypertension: renin measurements to indicate hypersecretion and contralateral suppression, estimate renal plasma flow and score for surgical durability. *Amer. J. Med.* **55**, 402–414

(170) Brown, J. J., Fraser, R., Lever, A. F., Morton, J. J., Robertson, J. I. S., Tree, M., Bell, P. R. F., Davidson, J. K. & Ruthven, I. S. (1973) Hypertension and secondary hyperaldosteronism associated with a renin-secreting renal juxtaglomerular cell tumour. *Lancet* **ii**, 1228–1232

(171) Schambelan, M., Howes, E. L., Stockigt, J. R., Noakes, C. A. & Biglieri, E. G. (1973) Role of renin and aldosterone in hypertension due to a renin-secreting tumour. *Amer. J. Med.* **55**, 86–92

(172) Robertson, P. W., Klidjian, A., Harding, L. K., Walters, G., Lee, M. R. & Robb-Smith, A. H. T. (1967) Hypertension due to a renin-secreting renal tumour. *Amer. J. Med.* **43**, 963–976

(173) Sealey, J. E., Buhler, F. R., Laragh, J. H. & Vaughan, E. D. (1973) The physiology of renin secretion in essential hypertension. Estimation of renin secretion rate and renal plasma flow from peripheral and renal vein renin levels. *Amer. J. Med.* **55**, 391–401

(174) Ledingham, J. M. (1971) Blood pressure regulation in renal failure. *J. Roy. Coll. Physns.* **5**, 103–134

(175) Brown, J. J., Curtis, J. R., Lever, A. F., Robertson, J. I. S., de Wardener, H. E. & Wing, A. J. (1969) Plasma renin concentration and the control of blood pressure in patients on maintenance haemodialysis. *Nephron* **6**, 329–349

(176) Medina, A., Bell, P. R. F., Briggs, J. D., Brown, J. J., Fine, A., Lever, A. F., Morton, J. J., Paton, A. M., Robertson, J. I. S., Tree, M., Waite, M. A., Weir, R. J. & Winchester, J. (1972) Changes of blood pressure, renin and angiotensin after bilateral nephrectomy in patients with chronic renal failure. *Brit. Med. J.* **4**, 694–696

(177) Giese, J. (1973) Renin, angiotensin and hypertensive vascular damage: a review. *Amer. J. Med.* **55**, 315–332

(178) Gavras, H., Brown, W. C. B., Brown, J. J., Lever, A. F., Linton, A. L., Macadam, R. F., McNicol, G. P., Robertson, J. I. S. & Wardrop, C. (1971) Microangiopathic haemolytic anaemia and the development of the malignant phase of hypertension. *Circ. Res.* **28**, *Suppl.* **2**, 127–142

(179) Weir, R. J., Briggs, E., Mack, A., Taylor, L., Browning, J., Naismith, L. & Wilson, E. (1971) Blood pressure in women after one year of oral contraception. *Lancet* **1**, 467–471

(180) Weir, R. K., Tree, M. & McElwee, G. (1974) Changes in blood pressure and in plasma renin, renin-substrate and angiotensin II concentrations in women taking contraceptive steroids. *Proc. Int. Congr. Endocrinol. 4th*, pp. 1021–1025, Excerpta Medica, Amsterdam

(181) Laragh, J. H., Sealey, J. E. & Ledingham, J. G. G. (1967) Oral contraceptives: renin, aldosterone and high blood pressure. *J. Amer. Med. Ass.* **201**, 918–922

(182) Woods, J. W. (1967) Oral contraceptives and hypertension. *Lancet* **2**, 653–654

(183) Skinner, S. L., Lumbers, E. R. & Symonds, E. M. (1969) Alteration by oral contraceptives of normal menstrual changes in plasma renin activity, concentration and substrate. *Clin. Sci.* **36**, 67–76

(184) Cain, M. D., Walters, W. A. & Catt, K. J. (1971) Effects of oral contraceptive therapy on the renin–angiotensin system. *J. Clin. Endocrinol.* **33**, 671–676

(185) Atkinson, A. B. & Kellett, R. J. (1974) Value of intravenous urography in investigating hypertension. *J. Roy. Coll. Phys.* (*London*) **8**, 175–181

(186) Miller, N. E., Dicara, L. V., Solomon, H., Weiss, J. M. & Dworkin, B. (1970) Learned modifications of autonomic functions: a review of some new data. *Circ. Res.* **26**, *Suppl.* **1**, 3–11

(187) DeQuattro, V. & Miura, Y. (1973) Neurogenic factors in human hypertension: mechanism or myth? *Amer. J. Med.* **55**, 362–378

(188) Folkow, B. (1971) The haemodynamic consequences of adaptive structural changes of the resistance vessels in hypertension. *Clin. Sci.* **41**, 1 12

(189) Ledingham, J. M. (1971) Mechanisms in renal hypertension. *Proc. Roy. Soc. Med.* **64**, 409–418

(190) Guyton, A. C., Coleman, T. G., Cowley, A. W., Scheel, K. W., Manning, R. D. & Norman, R. A. (1972) Arterial pressure regulation. *Amer. J. Med.* **52**, 584–594
(191) Dahl, L. K., Knudsen, K. D., Heine, M. A. & Leitl, G. J. (1968) Effects of chronic excess salt ingestion. *Circ. Res.* **22**, 11–18
(192) Douglas, B. H., Guyton, A. C., Langston, J. B. & Bishop, V. S. (1964) Hypertension caused by salt loading. II. Fluid volume and tissue pressure changes. *Amer. J. Physiol.* **207**, 669–671
(193) Dusting, G. J., Harris, G. S. & Rand, M. J. (1973) A specific increase in cardiovascular reactivity related to sodium retention in DOCA-salt-treated rats. *Clin. Sci. Mol. Med.* **45**, 571–581
(194) Davies, D. L., Beevers, D. G., Brown, J. J., Fraser, R., Ferrriss, J. B., Lever, A. F., Medina, A., Morton, J. J. & Robertson, J. I. S. (1974) Sodium and the renin–angiotensin system in patients with hypertension. *Proc. Int. Congr. Endocrinol. 4th*, pp. 693–698, Excerpta Medica, Amsterdam
(195) Editorial (1974) Volume-dependent essential hypertension. *Brit. Med. J.* **1**, 470
(196) Schalekamp, M. A., Beevers, D. G., Briggs, J. D., Brown, J. J., Davies, D. L., Fraser, R., Lebel, M., Lever, A. F., Medina, A., Morton, J. J., Robertson, J. I. S. & Tree, M. (1973) Hypertension in chronic renal failure: an abnormal relation between sodium and the renin–angiotensin system. *Amer. J. Med.* **45**, 379–390
(197) Blumberg, A., Hegstrom, R. M., Nelp, W. B. & Schribner, B. H. (1967) Extracellular volume in patients with chronic renal disease treated for hypertension by sodium restriction. *Lancet* **ii**, 69–73
(198) Bianchi, G., Ponticelli, C., Bardi, U., Redaelli, B., Campolo, L., de Ponti, C. & Graziani, G. (1972) Role of the kidney in salt- and water-dependent hypertension of end-stage renal disease. *Clin. Sci.* **42**, 47–55
(199) Dathan, J. R. E., Johnson, D. B. & Goodwin, F. J. (1973) The relationship between body fluid compartment volumes, renin activity and blood pressure in chronic renal failure. *Clin. Sci. Mol. Med.* **45**, 77–78
(200) Thompson, J. M. A. & Dickinson, C. J. (1973) Relation between pressure and sodium excretion in perfused kidneys from rabbits with experimental hypertension. *Lancet* **ii**, 1362–1363
(201) Johnson, J. A. & Davis, J. O. (1973) Angiotensin II. Important role in the maintenance of arterial blood pressure. *Science* **179**, 906–907
(202) Strewler, G. J., Hinrichs, K. J., Guiod, L. R. & Hollenberg, N. K. (1972) Sodium intake and vascular smooth muscle responsiveness to norepinephrine and angiotensin in the rabbit. *Circ. Res.* **31**, 758–766
(203) Davies, D. L., Schalekamp, M. A., Beevers, D. G., Brown, J. J., Briggs, J. D., Lever, A. F., Medina, A. M., Molton, J. J., Robertson, J. I. S. & Tree, M. (1973) Abnormal relation between exchangeable sodium and the renin–angiotensin system in malignant hypertension and in hypertension with chronic renal failure. *Lancet* **i**, 683–687
(204) Warren, D. J. & Ferriss, T. F. (1970) Renin secretion in renal hypertension. *Lancet* **i**, 159–163
(205) Gavras, H., Brunner, H. R., Vaughan, E. D. & Laragh, J. H. (1973) Angiotensin–sodium interaction in blood pressure maintenance of renal hypertensive and normotensive rats. *Science* **180**, 1369–1372
(206) Lebel, M., Schalekamp, M. A., Beevers, D. G., Brown, J. J., Davies, D. L., Fraser, R., Kremer, D., Lever, A. F., Morton, J. J., Robertson, J. I. S., Tree, M. & Wilson, A. (1974) Sodium and the renin–angiotensin system in essential hypertension and mineralocorticoid excess. *Lancet* **ii**, 308–310
(207) Crane, M. G. & Harris, J. J. (1970) Effect of spironolactone in hypertensive patients *Amer. J. Med. Sci.* **260**, 311–330
(208) Spark, R. F. & Melby, J. C. (1971) Hypertension and low plasma renin activity: presumptive evidence for mineralocorticoid excess. *Ann. Intern. Med.* **75**, 831–836
(209) Carey, R. M., Douglas, J. G., Schweikert, J. R. & Liddle, G. W. (1972) The syndrome of essential hypertension and suppressed plasma renin activity. *Arch. Intern. Med.* **130**, 849–854

(210) Woods, J. W., Chapel-Hill, N. C., Liddle, G. W., Stant, E. G., Michelakis, A. M. & Brill, A. B. (1969) Effect of an adrenal inhibitor in hypertensive patients with suppressed renin. *Arch. Intern. Med.* **123**, 366–370
(211) Dickinson, C. J. (1965) *Neurogenic Hypertension* Blackwell, Oxford
(212) Tarazi, R. C. & Dustan, H. P. (1973) Neurogenic participation in essential and renovascular hypertension assessed by acute ganglionic blockade: correlation with haemodynamic indices and intravascular volume. *Clin. Sci.* **44**, 197–212
(213) Esler, M. D. & Nestel, P. J. (1973) Renin and sympathetic nervous system responsiveness to adrenergic stimuli in essential hypertension. *Amer. J. Cardiol.* **32**, 643–649
(214) Louis, W. J., Doyle, A. E. & Anavekar, S. (1973) Plasma noradrenaline in essential hypertension. *N. Engl. J. Med.* **288**, 599–601
(215) Lewis, G. P. & Reit, E. (1965) The action of angiotensin and bradykinin on the superior cervical ganglion of the cat. *J. Physiol.* **179**, 538–553
(216) Rosendorff, C., Lowe, R. D., Lavery, H. & Cranston, W. I. (1970) Cardiovascular effects of angiotensin mediated by the central nervous system in the rabbit. *Cardiovasc. Res.* **4**, 36–43
(217) Ferrario, C. M., Gildenberg, P. L. & McCubbin, J. W. (1972) Cardiovascular effects of angiotensin mediated by the central nervous system. *Circ. Res.* **30**, 257–262
(218) Ferrario, C. M., Dickinson, C. J. & McCubbin, J. W. (1970) Central vasomotor stimulation by angiotensin. *Clin. Sci.* **39**, 239–245
(219) Scroop, G. C. & Lowe, R. D. (1969) Efferent pathways of the cardiovascular response to vertebral artery infusions of angiotensin in the dog. *Clin. Sci.* **37**, 605–619
(220) Joy, M. D. & Lowe, R. D. (1970) The site of cardiovascular action of angiotensin II in the brain. *Clin. Sci.* **39**, 327–336
(221) Joy, M. D. (1971) The intramedullary connections of the area postrema involved in the central cardiovascular response to angiotensin II. *Clin. Sci.* **41**, 89–100
(222) Brenner, B. M., Troy, J. L., Daugharty, T. M., Deen, W. M. & Robertson, C. R. (1972) Dynamics of glomerular ultrafiltration in the rat. II. Plasma flow dependence of GFR. *Amer. J. Physiol.* **223**, 1184–1190
(223) Friedman, M., Selzer, A. & Rosenblum, H. (1941) The renal blood flow in hypertension as determined in patients with variable, with early and with long-standing hypertension. *J. Amer. Med. Ass.* **117**, 92–95
(224) Goldring, W., Chasis, H., Ranges, H. A. & Smith, H. W. (1941) Effective renal blood flow in subjects with essential hypertension. *J. Clin. Invest.* **20**, 637–653
(225) Schalekamp, M. A. D. H., Schalekamp-Kuyken, M. P. A. & Birkenhager, W. H. (1970) Abnormal renal haemodynamics and renin suppression in hypertensive patients. *Clin. Sci.* **38**, 101–110
(226) Schalekamp, M. A. D. H., Krauss, X. H., Kolsters, G., Schalekamp, M. P. A. & Birkenhager, W. H. (1973) Renin suppression in hypertension in relation to body fluid volumes, patterns of sodium excretion and renal haemodynamics. *Clin. Sci. Mol. Med.* **45**, *Suppl.* **1**, 283s–286s
(227) Windhager, E. E., Lewy, J. E. & Spitzer, A. (1969) Intrarenal control of proximal tubular reabsorption of sodium and water. *Nephron* **6**, 247–259
(228) Bank, N., Aynedjian, H. S. & Wada, T. (1972) Effect of peritubular capillary perfusion rate on proximal sodium reabsorption. *Kidney Int.* **1**, 397–405
(229) Brenner, B. M. & Galla, J. H. (1971) Influence of postglomerular hematocrit and protein concentration on rat nephron fluid transfer. *Amer. J. Physiol.* **220**, 148–161
(230) Falchuk, K. H., Brenner, B. M., Tadokoro, M. & Berliner, R. W. (1971) Oncotic and hydrostatic pressures in peritubular capillaries and fluid reabsorption by proximal tubule. *Amer. J. Physiol.* **220**, 1427–1433
(231) Lowenstein, J., Beranbaum, E. R., Chasis, H. & Baldwin, D. S. (1970) Intrarenal pressure and exaggerated natriuresis in essential hypertension. *Clin. Sci.* **38**, 359–374
(232) Schalekamp, M. A. D. H., Kolsters, G., Birkenhager, W. H. & Lever, A. F. (1974) Pathogenetic aspects of low-renin hypertension. In *Hypertension: Current Problems* (Distler, A. & Wolff, H. P.), pp. 133–142, Thieme Verlag, Stuttgart

(233) Floyer, M. A. (1957) Role of the kidney in experimental hypertension. *Brit. Med. Bull.* **13**, 29–32

(234) Beilin, L. J. & Ziakas, G. (1972) Vascular reactivity in post-deoxycorticosterone hypertension in rats and its relation to 'irreversible' hypertension in man. *Clin. Sci.* **42**, 579–590

(235) Kramer, P. & Ochwadt, B. (1972) Sodium excretion in Goldblatt hypertension. Long-term separate kidney function studies in rats by means of a new technique. *Pflugers Arch.* **332**, 332–345

(236) Gross, F., Regoli, D. & Schaechtelin, G. (1963) Renal content and blood concentration of renin. *Mem. Soc. Endocrinol.* **13**, 293–300

(237) Schalekamp, M. A. D. H., Krauss, X. H., Schalekamp-Kuyken, M. P. A., Kolsters, G. & Birkenhager, W. H. (1971) Studies on the mechanism of hypernatriuresis in essential hypertension in relation to measurements of plasma renin concentration, body fluid compartments and renal function. *Clin. Sci.* **41**, 219–231

(238) Appleby, J. H., Gibson, G., Norymberski, J. K. & Stubbs, R. D. (1954) The use of sodium borohydride in the determination of 17-hydroxylated steroids. *Biochem. J.* **57**, xiv–xv

(239) Cope, C. L. & Black, E. G. (1958) The behaviour of [^{14}C]cortisol and estimation of cortisol production rate in man. *Clin. Sci.* **17**, 147–163

(240) Mattingly, D. (1962) A simple fluorimetric method for the estimation of free 11-hydroxycorticoids in human plasma. *J. Clin. Pathol.* **15**, 374–379

(241) Murphy, B. E. P. (1969) Protein binding and the assay of nonantigenic hormones. *Recent Progr. Horm. Res.* **25**, 563–601

(242) Silber, R. H. & Porter, C. C. (1954) The determination of 17,21-dihydroxy-20-ketosteroids in urine and plasma. *J. Biol. Chem.* **210**, 923–932

(243) Norymberski, J. K. (1952) The determination of urinary corticosteroids. *Nature* (*London*) **170**, 1074–1075

(244) Beardwell, C. G., Burke, C. W. & Cope, C. L. (1968) Urinary free cortisol measured by competitive protein binding. *J. Endocrinol.* **42**, 79–89

(245) Mason, P. A. & Fraser, R. (1974) Simultaneous estimation of aldosterone, 11-deoxycorticosterone, 18-hydroxy-11-deoxycorticosterone, corticosterone, cortisol and 11 deoxycortisol in human plasma by gas–liquid chromatography with electron capture detection. *J. Endocrinol.* in the press

(246) Fraser, R. & James, V. H. T. (1968) Double isotope assay of aldosterone corticosterone and cortisol in human peripheral plasma. *J. Endocrinol.* **40**, 59–72

(247) Schalekamp, M. A., Beevers, D. G., Kolsters, G., Lebel, M., Fraser, R. & Birkenhäger, W. H. (1974) Body fluid volume in low-renin hypertension. *Lancet* **ii**, 310–311

(248) Brown, J. J., Lever, A. F., Robertson, J. I. S. & Schalekamp, M. A. (1974) Renal abnormality of essential hypertension. *Lancet* **ii**, 320–323

(249) Semple, P. F. & Morton, J. J. (1975) Angiotensin II and its heptapeptide and hexapeptide fragments in arterial and venous blood in man. *Clin. Sci. Mol. Med.* **48**, 2P

(250) Semple, P. F., Boyd, A. S., Dawes, P. M. & Morton, J. J. (1975) Angiotensin II and its heptapeptide (2–8), hexapeptide (3–8) and pentapeptide fragments in arterial and venous blood of man. Submitted for publication

(251) Semple, P. F. & Morton, J. J. (1975) Angiotensin II and angiotensin III in rat blood. *Eur. J. Clin. Invest.* in the press

(252) Cain, M. D., Catt, K. J. & Coghlan, J. P. (1970) Effect of circulatory fragments of angiotensin II on radioimmunoassay in arterial and venous blood. *J. Clin. Endocrinol.* **29**, 1639–1643

(253) Campbell, W. B., Brooks, S. N. & Pettinger, W. A. (1974) Angiotensin II- and angiotensin II-induced aldosterone release *in vivo* in the rat. *Science* **184**, 994–996

(254) Bangham, D. R., Robertson, I., Robertson, J. I. S., Robinson, C. J. & Tree, M. (1975) An international collaborative study of renin assay: introduction of an International Reference preparation of human renin. *Clin. Sci. Mol. Med.* in the press

Essays Med. Biochem. **1**, 59–80
Printed in Great Britain

Bile Acid Synthesis: An Alternative Pathway Leading to Hepatotoxic Compounds?

By I. W. PERCY-ROBB

University Department of Clinical Chemistry,
The Royal Infirmary, Edinburgh EH3 9YW, U.K.

Introduction

The injection of radioactively labelled cholesterol into laboratory animals (1) or human volunteers (2), or the addition of radioactively labelled cholesterol to the perfusate of an isolated liver perfusion preparation (3) leads to production of bile containing both radioactively labelled cholesterol and its acidic derivatives, the bile acids. The structural relationship between cholesterol and the bile acids was first demonstrated when a bile acid was produced by oxidation of the sterol, 5β-cholestane (coprostane), which had itself been prepared from cholesterol (4).

In mammals the major bile acids contain 24 carbon atoms and have the structures shown in Fig. 1. The two major bile acids synthesized by mammalian liver (the primary bile acids), namely cholic acid (3α,7α,12α-trihydroxy-5β-cholan-24-oic acid) and chenodeoxycholic acid (3α,7α-dihydroxy-5β-cholan-24-oic acid) differ only in the number of hydroxyl substituents present on the ring structure. The bile acids owe some of their unique physico-chemical properties to the presence of these hydroxyl substituents and to their close spatial relationship with one another which is a consequence of the *cis* orientation of the A and B rings.

Much work has been devoted, within recent years, to studies designed to establish the route by which cholesterol is converted into bile acids and to increase understanding of the mechanisms whereby the rate of the overall synthetic process is controlled. As a result it is now known that the major pathway of bile acid synthesis is initiated by modifications to the ring structure of the cholesterol molecule and that this is followed by cleavage of the side chain between the carbon atoms 24 and 25 (5). There is evidence, however, of an alternative pathway which is initiated by modifications to the cholesterol side chain (6) but whose quantitative significance is not known. This pathway leads predominantly to the formation of chenodeoxycholic rather than cholic acid and may also convert small amounts of cholesterol into lithocholic acid (3α-hydroxy-5β-cholan-24-oic acid) (6), which (Fig. 1) possesses toxic proper-

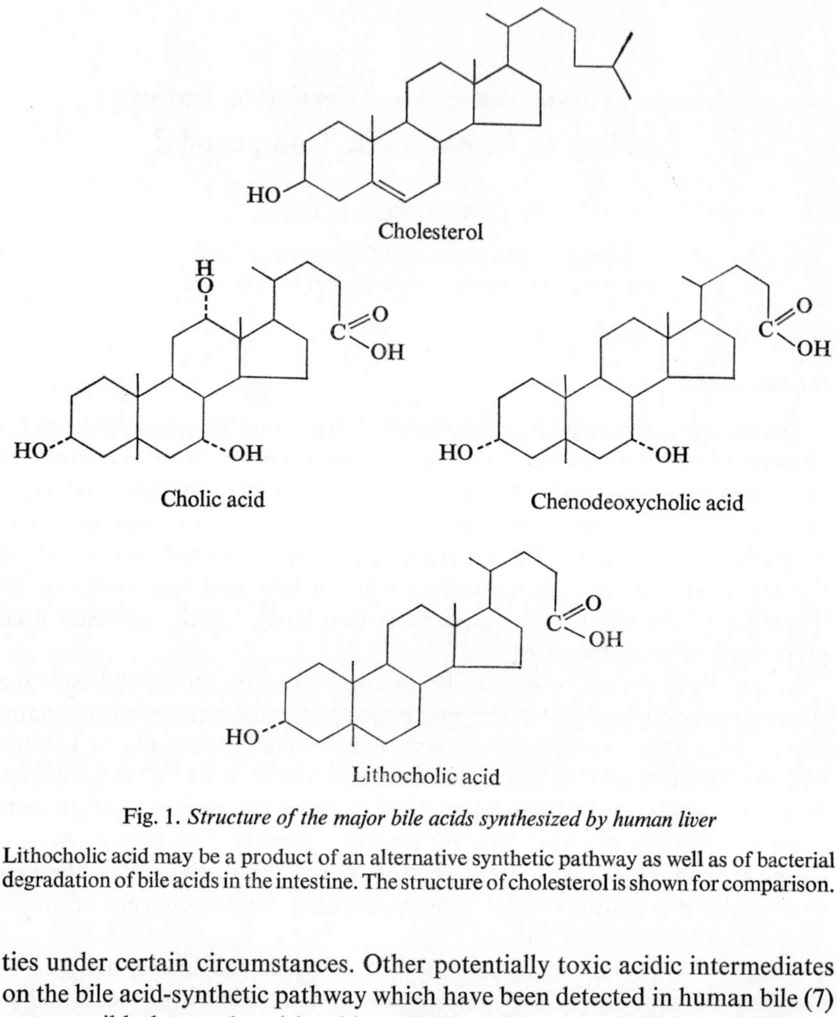

Cholesterol

Cholic acid

Chenodeoxycholic acid

Lithocholic acid

Fig. 1. *Structure of the major bile acids synthesized by human liver*

Lithocholic acid may be a product of an alternative synthetic pathway as well as of bacterial degradation of bile acids in the intestine. The structure of cholesterol is shown for comparison.

ties under certain circumstances. Other potentially toxic acidic intermediates on the bile acid-synthetic pathway which have been detected in human bile (7) may possibly be produced by this pathway. The existence of this alternative pathway suggests that (a) the liver may actually synthesize products that are themselves hepatotoxic and (b) under certain circumstances primary abnormalities of bile acid synthesis may cause liver disease.

The enterohepatic circulation of bile acids

Both cholic and chenodeoxycholic acids are conjugated in the liver with either glycine or taurine by the enzymic formation of an amide bond at position C-24. It is these conjugated acids that are secreted in bile. The entero-hepatic circulation consists of an efficient recycling system in which conjugated bile acids are reabsorbed largely from the terminal ileum, transported in the portal blood stream to the liver and rapidly resecreted in bile. This system conserves the bile acids which have been secreted in bile thus helping to maintain both the total bile acid pool and the concentration of bile acids secreted by the liver in bile.

The reabsorption of conjugated bile acids by the terminal ileum is performed by a system which transports cholic acid conjugates more rapidly than chenodeoxycholic acid derivatives, and taurine conjugates more rapidly than glycine conjugates (8). This system depends on an active transport mechanism, passive transport accounting for only a small proportion of the overall absorption (9). These absorption mechanisms are collectively responsible for the efficiency of the enterohepatic circulation of the bile acids. It has been estimated that 93–95% of the bile acids are reabsorbed during each cycle of the total bile acid pool and that the pool cycles four to eight times every 24h (10).

Intestinal bacteria have been shown to have profound effects on bile acid metabolism (11). Deconjugation (rupture of the amide bond) and dehydroxylation are the main bacterial transformations. These are performed mainly by the anaerobic bacterial species found in large numbers in the colon. The products of the dehydroxylation are the secondary bile acids deoxycholic acid (3α,12α-dihydroxy-5β-cholan-24-oic acid, formed by 7-dehydroxylation of cholic acid) and lithocholic acid (formed by 7-dehydroxylation of chenodeoxycholic acid). It appears that deoxycholic acid can be readily reabsorbed from the colon and that, thereafter, this secondary bile acid joins cholic acid and chenodeoxycholic acid in the enterohepatic circulation. On the other hand, lithocholic acid, which is relatively insoluble in water, is not readily reabsorbed from the colon and is largely excreted in the faeces.

The Clinical Importance of Bile Acids

The long-held clinical belief that bile is important in the maintenance of health, and that ill health may be associated with the production of 'bad' bile, has been justified in recent years by many studies which relate abnormalities of bile acid metabolism to clinical problems.

The bile acids, as a consequence of their physicochemical properties, have detergent-like actions which are physiologically important in solubilizing the products of the digestion of dietary fat (12, 13). They also play a role in solubilizing cholesterol in the bile (14–16). Loss of these functions may lead, in the case of the breakdown in solubilization of fatty acids in the upper small intestine, to malabsorption of dietary fat (17) or in the case of decreased solubilization of biliary cholesterol, to an increased tendency to formation of cholesterol gallstones (18). This may be due, in part, to a failure to secrete adequate concentrations of bile acids in the bile. The value of giving bile acids orally as therapeutic agents for dissolving cholesterol gallstones and thereby avoiding the necessity for biliary surgery, is currently being assessed (19, 20).

The synthesis of bile acids represents quantitatively the major pathway by which cholesterol is degraded and excreted from the body. The introduction into clinical medicine of cholestyramine, an anion-exchange resin possessing strong binding affinity for bile acids in the intestinal lumen (21) and therefore capable of preventing their normal reabsorption, has far-reaching implications in that the rate of bile acid excretion in the faeces can now be controlled. Because of the binding of bile acids to cholestyramine within the intestine, bile acid synthesis, and hence cholesterol degradation, is increased and may lead to a fall in the plasma cholesterol concentration. This may be of benefit to patients with hypercholesterolaemia and atherosclerosis (22).

Some toxic effects of lithocholic acid

The evidence relating bile acids to liver injury and liver disease has been extensively reviewed (23). Lithocholic acid may cause both long-term and short-term adverse effects on the structure and function of the liver as well as producing immediate systemic effects which include the production in man, after intramuscular injection, of pyrogenic reactions. Lysis of human erythrocytes is produced *in vitro* (24).

Holsti (25), in 1960, reported development of cirrhosis of the liver in rabbits given daily instillations of lithocholic acid into the stomach. This appeared to be dose-related. Abnormalities of liver structure have since been demonstrated in a variety of mammalian species after oral administration of either lithocholic acid itself or its glycine or taurine conjugates.

Recognition of the toxic effects of lithocholic acid led to further studies of its metabolism and to the identification, in human bile, of sulphates of the conjugated acid (26, 27). This represents a new pathway of bile acid metabolism which accounts for about 50% of the total lithocholic acid in human bile. It has been shown in the guinea pig that intestinal absorption of the sulphate esters of conjugated lithocholic acid is low when compared with the non-sulphated compounds (28) which provides support for the concept that sulphation of the toxic lithocholic acid enhances its faecal elimination.

The bile acids possess choleretic properties and are now generally accepted as major determinants of bile flow (29). Both lithocholic acid and its taurine conjugate are rapidly excreted in the bile after their intravenous infusion in the rat (30). The appearance of lithocholate in the bile decreases the bile flow rate when its molar excretion rate exceeds the combined excretion rates of cholic acid and chenodeoxycholic acid. This appears to be a consequence of mechanical obstruction to the biliary tree by precipitated lithocholate. It is unlikely that the cholestatic effect of lithocholate has a direct clinical equivalent as only trace quantities of lithocholate, at most, have been observed in human bile (31).

Pathways of Bile Acid Synthesis

The conversion of cholesterol into bile acids involves a series of enzyme-mediated reactions which result in modification of the ring structure, in shortening of the cholesterol side chain and in oxidation of the carbon atom at position 24 to a carboxylic acid group.

Various experiments, including use of the isolated perfused liver preparation, have led to the conclusion that synthesis of bile acids occurs only in the liver (3, 32). Most of the studies which have led to the present understanding of the pathway of conversion were performed in tissues from rats; a comparatively small number have been performed in rhesus monkeys or human volunteers.

Overall conversion of cholesterol into bile acids is initiated by a series of reactions which occur in microsomal preparations and since these appear to be the analogue of the endoplasmic reticulum *in vitro*, it seems reasonable to assume that cholesterol → bile acid conversion *in vivo* commences in this portion of the hepatocyte. The initial steps in bile acid synthesis *in vitro* are followed by reactions that take place in the supernatant fraction. Final conversion occurs in the mitochondrial fraction.

Synthesis of cholic acid

(i) *Microsomal reactions* (*Fig*. 2). The first step in the conversion of cholesterol into bile acids is generally considered to be the formation of 7α-hydroxycholesterol. This has largely been deduced from the studies of Bergström (33) and of Eriksson (34) in the bile fistula rat. It has also been demonstrated that 7α-hydroxycholesterol is rapidly converted into bile acids in experimental animals (35) and human volunteers (36).

The cholesterol 7α-hydroxylase system is present in the microsomal fraction of the liver and requires oxygen, NADPH and a thiol cofactor (37, 38). Spontaneous autoxidation of cholesterol *in vitro* leads to the formation of several products, including 7α-hydroxycholesterol, unless a radical scavenger such as cysteamine is added to the liver microsomal preparation (Fig. 3). The 7α-hydroxylase reaction is cytochrome *P*-450-dependent and is inhibited by CO.

Further metabolism of 7α-hydroxycholesterol takes place in microsomal preparations in the presence of NAD^+. It is first rapidly converted into the 3-oxo compound which is in turn changed, by enzymic migration of the Δ^{5-6} double bond to the 4-5 position into 7α-hydroxycholest-4-en-3-one (39–41), a compound which can be metabolized *in vivo* with the formation of

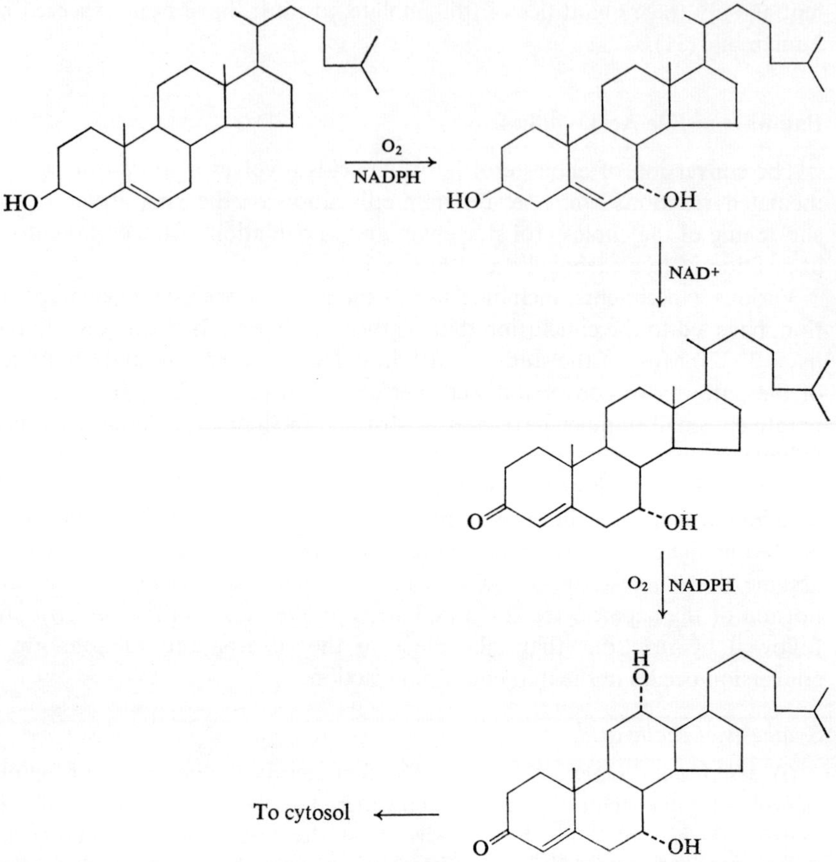

Fig. 2. *Outline of the first steps (microsomal reactions) in the conversion of cholesterol into the bile acids*

both cholic acid and chenodeoxycholic acid. Further metabolism of 7α-hydroxycholest-4-en-3-one can take place in microsomal preparations in the presence of oxygen and NADPH and under these circumstances a hydroxyl group is inserted into the 12α position with the formation of 7α,12α-dihydroxycholest-4-en-3-one (42, 43).

(ii) *Reactions in the liver supernatant fraction* (*Fig.* 4). Further metabolism of 7α,12α-dihydroxycholest-4-en-3-one appears to take place in the supernatant fraction of liver suggesting that it must first be removed from the endoplasmic reticulum. This hypothesis requires the assumption that the supernatant fraction truly represents an environment within the whole hepatocyte which is unrelated to either the endoplasmic reticulum or the mitochondria.

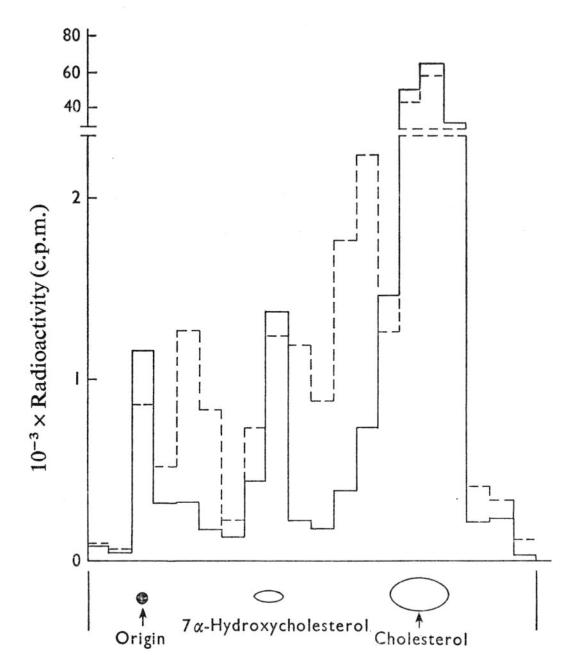

Fig. 3. *Conversion of radioactively labelled cholesterol into 7α-hydroxycholesterol in the presence of rat liver microsomal fraction, NADPH and oxygen*

In the presence of cysteamine (0.1 mol·l^{-1}) (——) only 7α-hydroxycholesterol appears as product whereas without cysteamine (----) several products, both less polar and more polar, are found.

Saturation of the Δ^4 double bond and reduction of the 3-oxo group of 7α,12α-dihydroxycholest-4-en-3-one occurs very efficiently in the liver supernatant fraction with the formation of 3α,7α,12α-trihydroxy-5β-cholestane (44). The enzyme systems involved in this transformation have been partially purified. Their cofactor is NADPH.

Current evidence suggests that the hydrogen atoms added in the 5β and 3β positions are derived from NADPH whereas that added at position C-4 comes from the medium (45). The saturation of the double bond in the unsaturated ketone is highly stereospecific for the insertion of the hydrogen into the 5β position. This contrasts with the major pathway for the hepatic metabolism of cortisol in which the 5α position is preferred. The two reactions appear to be promoted by different enzyme systems.

(iii) *Reactions in liver mitochondrial fraction* (*Fig.* 5). The injection of [26-^{14}C]cholesterol into experimental animals leads to $^{14}CO_2$ excretion in the expired air (46). A similar process can be shown to occur in liver mitochondrial fractions (47, 48).

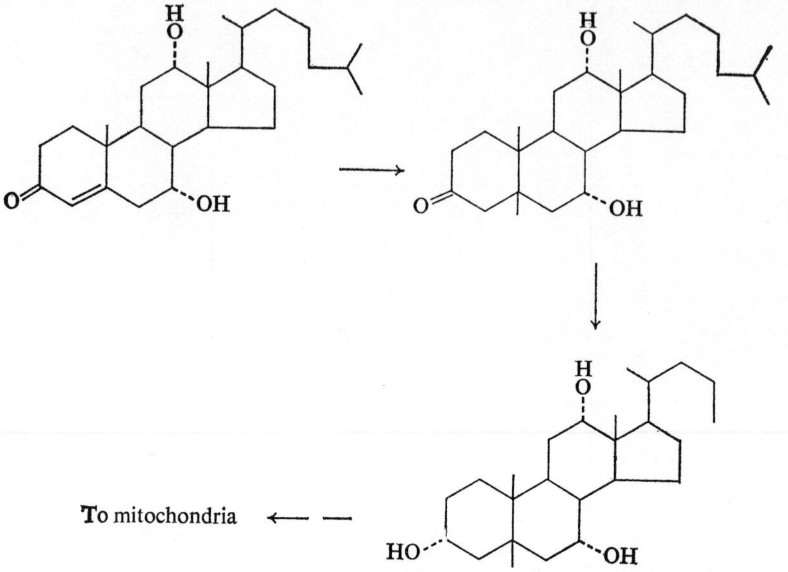

To mitochondria

Fig. 4. *Reactions occurring in the rat liver supernatant (post-*105 000 **g** *for* 60 *min) fraction*

Modification of the side chain of 3α,7α,12α-trihydroxy-5β-cholestane produced in the supernatant fraction appears to be initiated by 26-hydroxylation. This reaction occurs in the isolated mitochondrial fraction (49) but can also occur in the microsomal fraction (50). However, whereas the mitochondrial fraction is highly specific for hydroxylation at position C-26, the microsomal enzyme system is much less specific, being rather more active for adjacent positions. Apparently, therefore, the 26-hydroxylase system in the bile acid synthesis pathway is located within the mitochondria and, since it is inhibited by CO, is presumably cytochrome *P*-450-dependent.

The 3α,7α,12α,26-tetrahydroxy-5β-cholestane produced by 26-hydroxylation can be further oxidized in mitochondria to the corresponding 3α,7α,12α-trihydroxy-5β-cholestanoic acid. This can, in turn, be still further metabolized in the mitochondria with the formation of propionyl-CoA and cholyl-CoA (51).

That the final events in side chain cleavage involve a β-oxidation mechanism is strongly suggested by the isolation of 3α,7α,12α,24-tetrahydroxycholestanoic acid from the rat liver mitochondrial preparations and by the demonstration that this tetrol can be converted *in vivo* into cholic acid (52).

Synthesis of chenodeoxycholic acid

Cholic acid is not converted *in vivo* into chenodeoxycholic acid and the converse, though possible, is not a reaction of major quantitative significance.

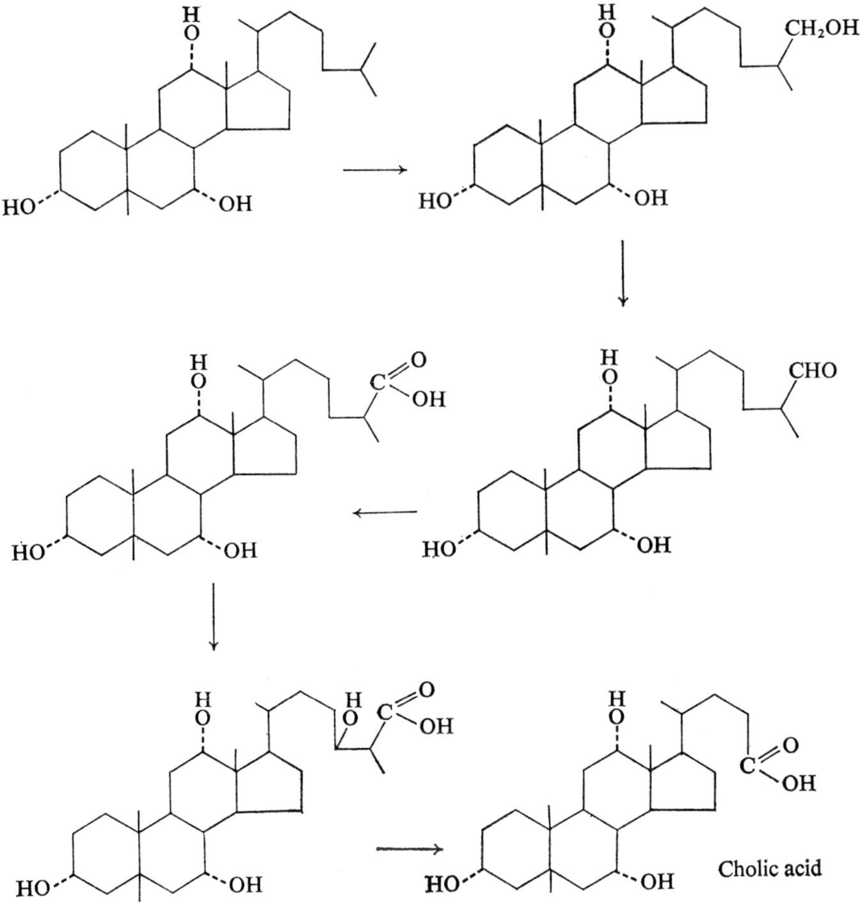

Cholic acid

Fig. 5. *Side chain cleavage in the biosynthesis of bile acids*

Apparently insertion of a 12α-hydroxyl group does not proceed smoothly once side chain hydroxylation of cholesterol derivatives has been initiated (36, 37).

Since 7α-hydroxycholesterol gives rise to both cholic acid and chenodeoxycholic acid (36), synthesis of both acids would appear to be initiated by a common series of reactions. A branch-point (Fig. 6) must occur, however, at which an intermediate metabolite is directed towards the synthesis of either cholic acid or chenodeoxycholic acid.

Various studies have shown that the microsomal 12α-hydroxylase enzyme system operates efficiently on 7α-hydroxycholest-4-en-3-one (41, 42). The

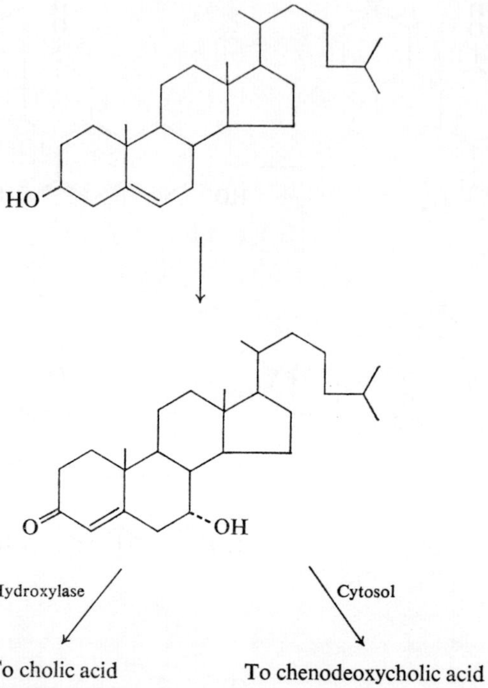

HO

O OH

12α-Hydroxylase Cytosol

To cholic acid To chenodeoxycholic acid

Fig. 6. *Branch-point in the synthesis of bile acids*

7α-Hydroxycholest-4-en-3-one is either subjected to 12α-hydroxylation in the microsomal fraction, with subsequent formation of cholic acid, or is immediately partitioned into the cytosol for conversion into chenodeoxycholic acid.

branch-point in the synthesis of bile acids consists, therefore, in the partitioning of this substrate within the hepatocyte either into the cytosol, where it undergoes further metabolism to chenodeoxycholic acid, or retention in the endoplasmic reticulum, where it is converted, by the microsomal 12α-hydroxylase system, into metabolites which lead finally to the formation of cholic acid.

The factors affecting the rate of 12α-hydroxylation of 7α-hydroxycholest-4-en-3-one and those which control its partitioning between the 12α-hydroxylase enzyme system in the microsomal fraction, on the one hand, and the enzymes in the liver cytosol, on the other, are still poorly understood. There are, however, situations in which the relative rates of synthesis of chenodeoxycholic acid and cholic acid are modified and these may give some information about the factors operative at the branch-point. These will be discussed below.

Alternative synthetic pathway

The possibility of an alternative pathway (Fig. 7) for the synthesis of bile

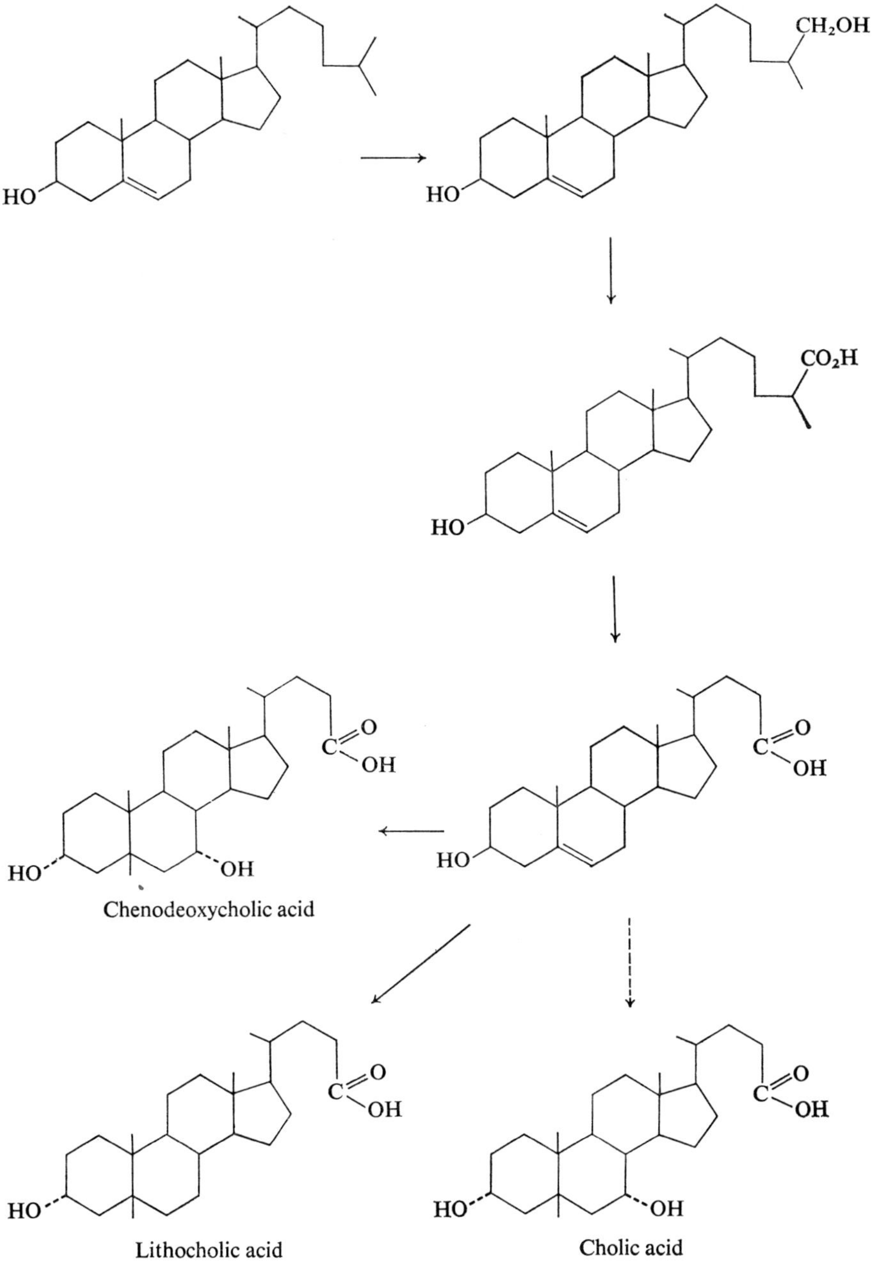

CH_2OH

CO_2H

Chenodeoxycholic acid

Lithocholic acid

Cholic acid

Fig. 7. *The proposed alternative pathway for the synthesis of bile acids from cholesterol*

The initial reactions lead to side chain cleavage followed by hydroxylation of the ring structure. This pathway leads primarily to the formation of chenodeoxycholic acid and lithocholic acid with the formation of minor amounts of cholic acid.

acids from cholesterol has been proposed on the basis of clinical studies and studies on liver preparations *in vitro*.

It has been shown experimentally that mitochondrial preparations from mouse and rat liver can hydroxylate cholesterol at C-26 in the side chain (48, 49) and although formation of cholic acid apparently does not proceed smoothly after this modification the possibility remains that the 26-hydroxycholesterol can act as an intermediate in bile acid formation. A comparison made between the metabolism of radioactively labelled 26-hydroxycholesterol and 7α-hydroxycholesterol in patients with biliary fistula showed that both compounds led to the appearance of radioactively labelled chenodeoxycholic acid and cholic acid in the bile (36). However, the incorporation of 7α-hydroxycholesterol was in proportion to the normal rate of synthesis of the two acids, whereas incorporation of 26-hydroxycholesterol was predominantly into chenodeoxycholic acid. 26-Hydroxycholesterol can also be shown to be converted into chenodeoxycholic acid and cholic acid in the rat and hamster (53).

Hanson (54) has shown that 3α,7α-dihydroxy-5β-cholestan-26-oic acid, a theoretical intermediate in an alternative synthetic pathway, can be recovered from human bile. Moreover, it is predominantly converted *in vivo* into chenodeoxycholic acid, only 2% of its radioactivity being recovered as cholic acid. These findings appear to confirm the presence of an alternative pathway initiated by side chain hydroxylation and leading predominantly to the formation of chenodeoxycholic acid. These studies too, confirm the apparent difficulty of inserting a 12α-hydroxyl group once side chain hydroxylation has occurred.

Additional evidence for some degree of side chain-initiated degradation of cholesterol comes from studies *in vitro* on preparations of rat liver mitochondria and supernatant fraction (55, 56) in which various polar steroids were formed including 3β-hydroxycholest-5-en-26-oic acid, 3β-hydroxy-5β-cholan-24-oic acid and lithocholic acid (3α-hydroxy-5β-cholan-24-oic acid). This suggests that mitochondrial preparations are capable not only of initiating hydroxylation of the cholesterol side chain but also of converting the side chain-modified cholesterol into bile acid. The latter component of the process requires epimerization of the 3β-hydroxyl group of the cholesterol molecule as well as saturation of the cholesterol double bond by insertion of a 5β-hydrogen. The mechanisms of these reactions have not been established.

Conjugates of cholic and chenodeoxycholic acids are the main bile acids present in the urine of infants with extrabiliary atresia and who have no enterohepatic circulation of bile acids. However, both 3β-hydroxy-5β-cholan-24-oic acid and 3α-hydroxy-5α-cholan-24-oic (allolithocholic) acid have also been recovered in small quantities (57). Since no enterohepatic circulation is present in these patients these bile acids could not have arisen from bacterial degradation in the intestine. Lithocholic acid was identified in the urine of one patient and also in the meconium (58).

Normally, side chain-hydroxylated derivatives of cholesterol are excreted in the bile in the form of sulphates (59). The appearance, in the urine of children with biliary atresia, of these and other derivatives of cholesterol, particularly allolithocholic and lithocholic acids, provide further evidence for the existence of minor alternative pathways of bile acid synthesis. There are, however, no data currently available which allow quantitative assessment of these pathways leading to the formation of lithocholic acid, other theoretical intermediates of bile acids and chenodeoxycholic acid itself. Nor is it possible clearly to distinguish chenodeoxycholic acid synthesized by this alternative pathway from that produced by the normal major pathway.

In the bile fistula rat up to 2% of an injected dose of radioactively labelled cholesterol is incorporated into bile acids within 5 h whereas with radioactively labelled 26-hydroxycholesterol up to 80% is incorporated within the same period (48). One interpretation of these results is that the pool size of 26-hydroxycholesterol, with which the radioactively labelled diol equilibrates, is very small and that 26-hydroxycholesterol is obligated to the pathway towards the bile acid synthesis. Alternatively, hydroxylation of cholesterol by the mitochondrial 26-hydroxylase system may represent a rate-controlling step in the alternative pathway of bile acid synthesis.

The Control of the Rate of Bile Acid Synthesis

There is good reason to believe that in the main pathway for bile acid synthesis the primary attack on the cholesterol molecule is insertion of an hydroxyl group at C-7, which is allylic to the cholesterol double bond, even though 12α-hydroxycholesterol can be converted *in vivo* into bile acids (60).

Prolonged biliary drainage has a profound effect on the rates of synthesis of both cholic acid and chenodeoxycholic acid (35, 61, 62), and since the branch-point for their synthesis lies in the insertion of an hydroxyl group at position C-12 it would seem likely that the rate-limiting step for the whole synthetic pathway would precede this stage.

Activity of the cholesterol 7α-hydroxylating enzyme system is affected by factors that modify the rate of conversion of cholesterol into bile acids (63, 64). Thus biliary drainage stimulates an increase in cholesterol 7α-hydroxylase activity which first becomes detectable about 18 h after drainage commences and reaches a maximum after 2–3 days (60). The time-course of the change in enzyme activity is similar to that which occurs in the overall rate of bile acid synthesis (35, 61) (Fig. 8). The activity of other enzymes concerned with the early stages of cholesterol metabolism does not change after biliary drainage (61).

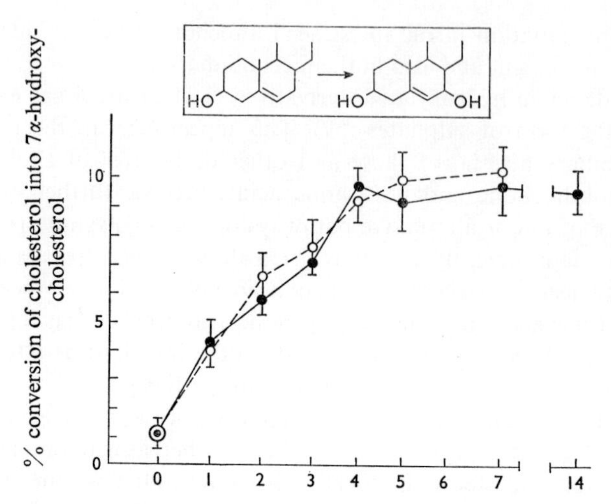

Fig. 8. *Pattern of induction of cholesterol 7α-hydroxylase enzyme system after interruption of the enterohepatic circulation of bile acids with the anion-binding agent cholestyramine or by total biliary drainage*

[Reproduced from ref. (60) with permission of the *American Journal of Medicine.*]

The suggestion that the activity of cholesterol 7α-hydroxylase is the major rate-determining step in overall conversion of cholesterol into bile acids is strengthened by the observation that biliary drainage increases incorporation of cholesterol or its precursor, mevalonate, into chenodeoxycholic acid and cholic acid. The increase in production of each acid is proportional to its rate of synthesis before commencing biliary drainage (3, 35) (Fig. 9). It is worthy of note that the overall conversion process can be stimulated to such an extent that the total amount of cholesterol degraded each day equals the sum of the plasma and liver cholesterol pools.

The effects of experimental biliary drainage have their clinical equivalent in the interruption of the enterohepatic circulation of bile acids which may be a result of resection or disease of the terminal ileum if sufficient to interfere with normal reabsorption of the bile acids (65). A controlled interruption of the enterohepatic circulation of the bile acids also occurs during therapeutic use of the anion-binding agent cholestyramine which binds bile acids and prevents their normal reabsorption. In both these clinical situations cholesterol degradation in the liver increases, and the rate of both cholic acid and chenodeoxycholic acid synthesis rises (66, 67). It is also clear that the rate of cholesterol synthesis in the liver rises in response to biliary drainage (62). Indeed it may be that the rate of cholesterol synthesis rises before detectable changes in bile acid synthesis occur.

Interruption of the enterohepatic circulation decreases the amount of bile

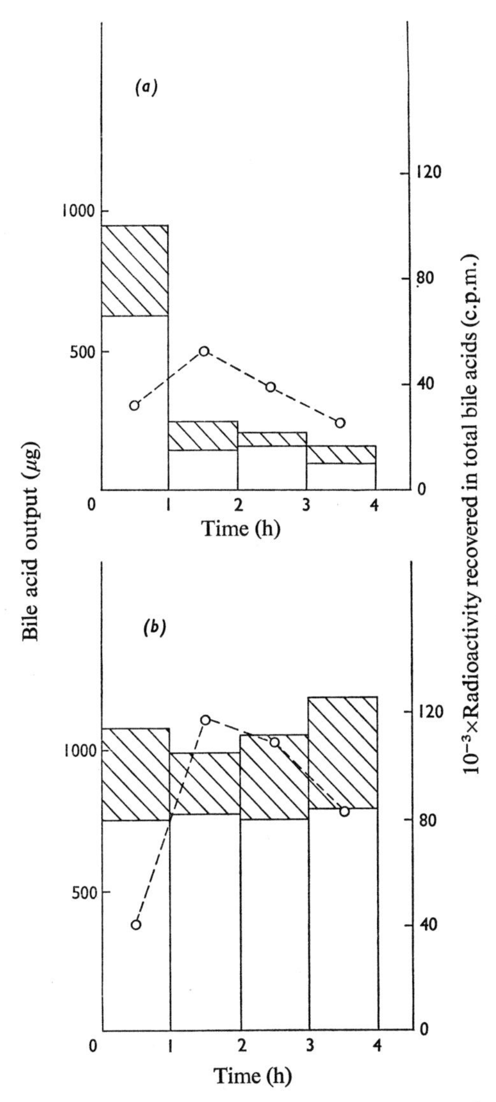

Fig. 9. *Change in bile acid production by isolated perfused rat liver after establishment of biliary fistula*

[Reproduced from ref. (3) with permission of the *Biochemical Journal*.] Secretion of bile acids and incorporation of [2-^{14}C]mevalonate into radiolabelled bile acids was studied in an isolated liver perfusion system. Livers were removed from donors immediately after the formation of biliary fistulae (*a*) and 40h after formation of biliary fistulae (*b*). ▧, Chenodeoxycholic acid; □, cholic acid. Radioactivity (○) represents incorporation into total bile acids.

acids returned to the liver in the portal venous blood stream. This is important since the amount of bile acids returning to the liver is held to be responsible for controlling, through a negative feed-back mechanism, the overall activity of the cholesterol 7α-hydroxylase enzyme system and hence the rate of endogenous bile acid synthesis. Studies by Bergström & Danielsson (68) showed that the intraduodenal infusion of chenodeoxycholic acid inhibited cholic acid synthesis. Their conclusions were confirmed by Mosbach and his colleagues (69) who demonstrated an inhibition of the induction of the cholesterol 7α-hydroxylase system which follows biliary drainage, when this was accompanied by intraduodenal infusion of chenodeoxycholic acid. There is, however, apparently an all-or-none effect of these infusions and no modulating effect on the activity of the enzyme system has so far been demonstrated.

The increase in hepatic cholesterol synthesis which accompanies bile diversion may, at first sight, appear to be an obvious response to an increase in the rate of cholesterol degradation. It should, however, be considered in the light of the well-established fact that both hepatic and plasma cholesterol serve as substrate for bile acid synthesis (3). Indeed, from a dynamic point of view, liver and plasma cholesterol can be considered as a single pool.

The pathway for the synthesis of cholesterol, and the mechanism whereby synthesis is regulated, have been reviewed (70). Acetyl-CoA is converted into β-hydroxy-β-methylglutaryl-CoA. This is reduced to mevalonate, six molecules of which are converted into squalene. Finally, cyclization of squalene forms lanosterol which is converted into cholesterol. It appears that although *in vitro* this reaction sequence can occur in most tissues from mammalian species, *in vivo* the liver and gastrointestinal tract are the most active.

The synthesis of cholesterol by the liver is regulated by a sensitive feed-back system, involving dietary cholesterol, which has been shown to inhibit the incorporation of acetate into cholesterol by liver slices (71). Conversion, by the β-hydroxy-β-methylglutaryl-CoA reductase enzyme, of the β-hydroxy-β-methylglutaryl-CoA intermediate into mevalonate is rate-limiting in the reaction sequence and is the step that is inhibited by dietary cholesterol supplements.

Hepatic cholesterol synthesis is also regulated by bile acids in the enterohepatic circulation. However, the mechanism of this effect is not well understood. Bile duct ligation, which causes a rise in the concentration of bile acids in the plasma and liver, leads to an increase in the rate of hepatic cholesterol synthesis, but the infusion of bile acids into the systemic circulation fails to inhibit hepatic cholesterol synthesis in the bile fistula animal (72). Thus various pieces of information together support the view that bile acids exert a regulatory effect on hepatic cholesterol synthesis by an indirect mechanism possibly through their well-established effect on intestinal cholesterol synthesis and transport. The apparent anomaly that there is a requirement for the synthesis of large amounts of two bile acids having similar physicochemical properties has been re-examined and in the rat a differential effect of the taurine conjugates of these acids has been proposed (73). Administration of taurocholate in the diet appears to inhibit the activity of hepatic β-hydroxy-β-methylglutaryl-CoA reductase and cholesterol 7α-hydroxylase enzymes whereas taurochenodeoxycholate infusion inhibits the β-hydroxy-β-methylglutaryl-CoA reductase but not the cholesterol 7α-hydroxylase. Whether these differential effects of conjugated bile acids on the activity of the cholesterol 7α-hydroxylase enzyme system are due to direct effects of the conjugates on the liver remains to be explored.

Altered relative rates of cholic acid and chenodeoxycholic acid synthesis

It is clear from what has already been said that the major pathway leading to formation of the bile acids includes a branch-point so that both cholic acid and chenodeoxycholic acid result from the initial 7α-hydroxylation of cholesterol. The possibility that chenodeoxycholic acid could be formed by an alternative route complicates interpretation of estimates of the rates at which cholic acid and chenodeoxycholic acid are formed, particularly with regard to rate-controlling factors. If the rate of synthesis of chenodeoxycholic acid rises, the increase could, from a theoretical standpoint, be due either to an increased flux of intermediates along the major degradative pathway leading to chenodeoxycholic acid, as well as cholic acid, or to an increase in the rate of chenodeoxycholic acid synthesis by way of the alternative pathway. There are certain experimental and clinical situations in which such an increase in the rate of synthesis of chenodeoxycholic acid does occur and which consequently provide an insight into the nature of these pathways.

Effects of thyroid hormones on bile acid synthesis. Thyroid-gland activity is known to affect cholesterol and bile acid metabolism both in experimental animals and human subjects (74, 75). Increased thyroid activity increases the rate of hepatic cholesterol synthesis and is accompanied by a fall in the plasma cholesterol concentration. Decreased thyroid activity, on the other hand, results in decreased hepatic cholesterol synthesis and an increase in plasma cholesterol. These changes may be due to the effects of thyroid hormone on the rate of bile acid or cholesterol synthesis or on cholesterol secretion in bile.

Thyroid-hormone treatment increases the rate and changes the pattern of total bile acid synthesis so that cholic acid and chenodeoxycholic acid are synthesized at more or less equal rates. The hypothyroid state causes little change in total bile acid synthesis but the proportion of chenodeoxycholic acid produced decreases slightly. These changes in relative bile acid synthesis rate are also evident when the response to biliary drainage is studied in the thyroid-hormone-treated rat. In this situation the expected rise in total bile acid synthesis rate occurs but the relative excretion rates of cholic acid and chenodeoxycholic acid remain unchanged (76, 77). These findings suggest that the ratio of cholic acid to chenodeoxycholic acid synthesis is controlled by factors which act on the major synthetic pathway, presumably at the 12α-hydroxylation stage. The alternative possibility that an increased rate of cholesterol side chain hydroxylation (26-hydroxylation) might be responsible for the increase in chenodeoxycholic acid synthesis seems unlikely since there is no evidence that biliary drainage has an effect on side chain hydroxylation.

Effect of liver disease on bile acid synthesis. It has been reported that liver disease is associated with abnormalities of bile acid metabolism. These include an increase in the concentration of bile acids in the plasma (79) and an increase in postprandial plasma bile acid concentrations in patients with cirrhosis (80), decreased clearance of bile acids from the blood and a change in the proportion of the various bile acids present in the plasma (81).

Carey (81) showed that the ratio of trihydroxylated bile acids to dihydroxylated bile acids (chenodeoxycholic acid and deoxycholic acid, 3α,12α-dihydroxycholan-24-oic acid) in plasma is decreased by hepatocellular injury. In advanced cirrhosis these changes can be shown to result largely from a decrease in the proportion of cholic acid with an increase in chenodeoxycholic acid (82).

The rates of synthesis of these two bile acids have been estimated in cirrhosis by the indirect isotope-dilution technique and that of cholic acid synthesis has been shown to be decreased whereas that of chenodeoxycholic acid is increased. There appears to be a correlation between the size of the change in synthesis rate of cholic acid and the severity of the cirrhosis (83).

These data may indicate that bile acid synthesis by the major pathway is inhibited and that the increased rate of chenodeoxycholic acid synthesis is due to alternative pathway activity or alternatively that the activity of the microsomal 12α-hydroxylase system is affected in liver disease. The latter explanation seems more likely but direct experimental evidence with which to test this theory is not available.

Bile acids in hyperlipoproteinaemia. The biochemical mechanisms responsible for hyperlipoproteinaemia are poorly understood. It has been claimed, for example, that bile acid synthesis rates are decreased in patients with primary hypercholesterolaemia (84) whereas in patients with combined hypercholesterolaemia and hypertriglyceridaemia they are said to be increased (85).

There is apparently a difference in the relative rates of synthesis of cholic acid and chenodeoxycholic acid in patients classified as suffering from Type II hyperlipoproteinaemia compared with normal subjects or those suffering from other forms of hyperlipoproteinaemia (86). In the Type II hyperlipoproteinaemic subjects the rate of cholic acid synthesis was 135 mg/24 h and that of chenodeoxycholic acid was 213 mg/24 h. This inversion of the normal relative rates of bile acid synthesis raises questions similar to those discussed in the section dealing with liver disease and for which there are still no direct data.

Conclusion

Remarkable advances in knowledge of the biology of the bile acids has resulted from integration of studies based largely on experiments *in vitro* (which have led to an understanding of the synthetic pathway and how it is controlled) with those based on clinical studies of liver and gastrointestinal disease and disorders of plasma cholesterol. The major synthetic pathway for bile acids is now well established, but understanding of the mechanisms responsible for controlling their rates of synthesis is still rudimentary. Little is known about the ways that bile acids gain access to the hepatocyte and hence to the major site of control of synthesis. The quantitative significance of the alternative pathway, and its possible role in producing toxic products which may cause or promote liver disease, remains an intriguing area of research for the future.

References

(1) Bloch, K., Berg, B. N. & Rittenberg, D. (1943) The biological conversion of cholesterol to cholic acid. *J. Biol. Chem.* **149**, 511–517

(2) Lewis, B. & Myant, N. B. (1967) Studies in the metabolism of cholesterol in subject with normal plasma cholesterol levels and in patients with essential hypercholesterolaemia. *Clin. Sci.* **32**, 201–213

(3) Percy-Robb, I. W. & Boyd, G. S. (1970) The synthesis of bile acids in perfused rat livers subjected to chronic biliary drainage. *Biochem. J.* **118**, 519–530

(4) Windaus, A. & Neukirchen, K. (1919) Die Umwandlung des Cholesterins in Cholansaure (28 Mitteilung über Cholesterin). *Ber. Deut. Chem. Ges.* **52**, 1915–1919

(5) Elliott, W. H. & Hyde, P. M. (1971) Metabolic pathways of bile acid synthesis. *Amer. J. Med.* **51**, 568–579

(6) Mitropoulos, K. A. & Myant, N. B. (1967) The formation of lithocholic acid, chenodeoxycholic acid and alpha- and beta-muricholic acids from cholesterol incubated with rat liver mitochondria. *Biochem. J.* **103**, 472–479.

(7) Hanson, R. F. & Williams, G. (1971) The isolation and identification of 3α,7α-dihydroxy-5β-cholestan-26-oic acid from human bile. *Biochem. J.* **121**, 863–864

(8) Schiff, E. R., Small, N. C. & Dietschy, J. M. (1972) Characterisation of the kinetics of the passive and active transport mechanisms for bile acid absorption in the small intestine and colon of the rat. *J. Clin. Invest.* **51**, 1351–1362

(9) Hepner, G. W., Hofmann, A. F. & Thomas, P. J. (1972) Metabolism of steroid and amino acid moieties of conjugated bile acids in man. *J. Clin. Invest.* **51**, 1889–1897

(10) Bunner, H., Hofmann, A. F. & Summerskill, W. H. J. (1972) Daily secretion of bile acids and cholesterol measured in health. *Gastroenterology* **62**, 188

(11) Hill, M. J. & Drasar, B. S. (1968) Degradation of bile salts by human intestinal bacteria. *Gut* **9**, 22–27

(12) Hofmann, A. F. & Borgström, B. (1962) Physico-chemical state of lipids in intestinal content during their digestion and absorption. *Fed. Proc. Fed. Amer. Soc. Exp. Biol.* **21**, 43–50

(13) Hofman, A. F. (1963) The functions of bile salts in fat absorption. The solvent properties of dilute micellar solutions of conjugated bile salts. *Biochem. J.* **89**, 57–68

(14) Dam, H. & Hegardt, F. G. (1971) The relation between and formation of gallstones rich in cholesterol and the solubility of cholesterol in aqueous solutions of bile salts and lecithin. *Z. Ernaehrungswiss.* **10**, 239–246
(15) Admirand, W. H. & Small, D. M. (1968) The physiochemical basis of cholesterol gallstone formation in man. *J. Clin. Invest.* **47**, 1043–1052
(16) Neiderhiser, D. H. & Roth, H. P. (1968) Cholesterol solubilisation by solutions of bile salts and bile salts plus lecithin. *Proc. Soc. Exp. Biol. Med.* **128**, 221–225
(17) Donaldson, R. M., Jr. (1965) Studies on the pathogenesis of steatorrhoea in the blind loop syndrome. *J. Clin. Invest.* **44**, 1815–1825
(18) Redinger, R. N. & Small, D. M. (1972) Bile composition, salt metabolism and gallstones. *Arch. Int. Med.* **130**, 618–631
(19) Danzinger, R. G., Hofmann, A. F. & Schoenfield, L. J. (1972) Dissolution of cholesterol gallstones by chenodeoxycholic acid. *N. Eng. J. Med.* **286**, 1–8
(20) Bell, G. D., Whitney, B. & Dowling, R. H. (1972) Gallstone dissolution in man using chenodeoxycholic acid. *Lancet* **ii**, 1213–1216
(21) Johns, W. H. & Bates, T. R. (1969) Quantification of the binding tendencies of cholestryamine I: Effect of structure and added electrolytes on the binding of unconjugated and conjugated bile-salt anions. *J. Pharm. Sci.* **58**, 179–183
(22) Grundy, S. H. (1972) Treatment of hypercholesterolemia by interference with bile acid metabolism. *Arch. Int. Med.* **130**, 638–648
(23) Palmer, R. H., Glickman, P. B. & Kappas, A. (1962) Pyrogenic and inflammatory properties of certain bile acids in man. *J. Clin. Invest.* **41**, 1573–1577
(24) Palmer, R. H. (1964) Haemolytic effect of steroids. *Nature* (*London*) **201**, 1134–1135
(25) Holsti, P. (1960) Cirrhosis of the liver induced in rabbits by gastric instillation of 3-monohydroxycholanic acid. *Nature* (*London*) **186**, 250
(26) Palmer, R. H. (1967) The formation of bile acid sulphates: a new pathway of bile acid metabolism in humans. *Proc. Nat. Acad. Sci. U.S.* **58**, 1047–1060
(27) Palmer, R. H. & Bolt, M. G. (1971) Bile acid sulphates. I. Synthesis of lithocholic acid sulphates and their identification in human bile. *J. Lipid Res.* **12**, 671–679
(28) Low-Beer, T. S., Tyor, M. P. & Lack, L. (1969) Effects of sulfation of taurolithocholic and glycolithocholic acids on their intestinal transport. *Gastroenterology* **56**, 721–726
(29) Wheeler, H. O. (1972) Secretion of bile acids by the liver and their role in the formation of hepatic bile. *Arch. Int. Med.* **130**, 533–541
(30) Javitt, N. B. & Emerman, S. (1968) Effect of sodium taurolithocholate on bile flow and bile acid excretion. *J. Clin. Invest.* **47**, 1001–1014
(31) Carey, J. B., Wilson, I. D., Zaki, F. G. & Hanson, R. F. (1966) The metabolism of bile acids with special reference to liver injury. *Medicine* (*Baltimore*) **45**, 461–470
(32) Harold, F. N., Felts, J. M. & Chaikoff, I. L. (1955) Fate of cholesterol-4-C^{14} and -26-C^{14} in the perfused liver. *Amer. J. Physiol.* **183**, 459–462
(33) Bergström, S. (1955) Formation and metabolism of bile acids. *Recent Chem. Progr.* **16**, 63–83
(34) Eriksson, S. (1957) Biliary excretion of bile acids and cholesterol in bile fistula rats. *Proc. Soc. Exp. Biol. Med.* **94**, 578–582
(35) Lindstedt, S. (1957). The formation of bile acids from 7α-hydroxycholesterol in the rat. *Acta Chem. Scand.* **11**, 417–420
(36) Anderson, K., Kok, E. & Javitt, N. B. (1972) Bile acid synthesis in man. Metabolism of 7α-hydroxycholesterol-^{14}C and 26-hydroxycholesterol-^{3}H. *J. Clin. Invest.* **51**, 112–117
(37) Scholan, N. & Boyd, G. S. (1968) The cholesterol 7α-hydroxylase enzyme system. *Hoppe-Seyler's Z. Physiol. Chem.* **349**, 1628–1630
(38) Grimwade, A. M., Lawson, M. E. & Boyd, G. S. (1973) Studies on rat-liver microsomal cholesterol 7α-hydroxylase. *Eur. J. Biochem.* **37**, 334–337
(39) Danielsson, H. (1961) On the metabolism of 7α-hydroxycholesterol in mouse liver homogenates. *Ark. Kemi* **17**, 363–367
(40) Hutton, H. R. B. & Boyd, G. S. (1966) The metabolism of cholest-5-ene-3β,7α-diol by rat-liver cell fractions. *Biochim. Biophys. Acta* **116**, 336–361

(41) Bjorkhem, I. & Danielsson, H. (1967) Formation and metabolism of some delta 4-cholestenols in rat-bile acids and steroids 185. *Eur. J. Biochem.* **2**, 403–413
(42) Danielsson, H. & Einarsson, K. (1966) On the conversion of cholesterol to 7α,12α-dihydroxycholest-4-ene-3-one. *J. Biol. Chem.* **241**, 1449–1454
(43) Einarsson, K. (1968). On the properties of the 12α-hydroxylase in cholic acid biosynthesis. *Eur. J. Biochem.* **5**, 101–108
(44) Berseus, O. & Bjorkhem, L. (1967). Enzymatic conversion of a delta-4-3-ketosteroid into a 3-alpha-hydroxy-5-beta steroid: mechanisms and stereochemistry of hydrogen transfer from NADPH. Bile acids and steroids 190. *Eur. J. Biochem.* **2**, 593–507
(45) Cronholm, T., Makino, I. & Sjövall, J. (1972) Steroid metabolism in rats given (1-$^2H_2$)ethanol. Biosynthesis of bile acids and reduction of 3-keto-5β-cholanoic acid. *Eur. J. Biochem.* **24**, 507–519
(46) Siperstein, M. D. & Chaikoff, J. L. (1952) C^{14}-cholesterol III. Excretion of carbons 4 and 26 in feces, urine and bile. *J. Biol. Chem.* **198**, 93–103
(47) Fredrickson, D. S. & Ono, K. (1956) The *in vitro* production of 25- and 26-hydroxycholesterol and their *in vivo* metabolism. *Biochim. Biophys. Acta* **22**, 183–184
(48) Berseus, O. (1965) On the stereospecificity of 26-hydroxylation of cholesterol. *Acta Chem. Scand.* **19**, 325–328
(49) Taniguchi, S., Hoshita, N. & Okuda, K. (1973) Enzymatic characteristics of CO-sensitive 26-hydroxylase system for 5α-cholestane-3α,7α,12α-triol in rat-liver mitochondria and its intramitochondrial localization. *Eur. J. Biochem.* **40**, 607–617
(50) Cronholm, T. & Johansson, G. (1970) Oxidation of 5β-cholestane-3α, 7α, 12α-triol by rat liver microsomes. *Eur. J. Biochem.* **16**, 373–381
(51) Suld, H. M., Staple, E. & Gurin, S. (1962). Mechanisms of formation of bile acids from cholesterol in oxidation of 5β-cholestane-3α,12α-triol and formation of propionic acid from the side-chain by rat liver mitochondria. *J. Biol. Chem.* **237**, 338–344
(52) Masui, T. & Staple, E. (1966) The formation of bile acids from cholesterol. The conversion of 5β-cholestane-3α,7α-triol-26-oic acid to cholic acid via 5β-cholestane-3α,7α,12α,24ξ-tetraol-26-oic acid by rat liver. *J. Biol. Chem.* **241**, 3889–3895
(53) Wachtel, N., Emerman, S. & Javitt, N. B. (1968) Metabolism of cholest-5-ene-3β,26-diol in the rat and hamster. *J. Biol. Chem.* **243**, 5207–5212
(54) Hanson, R. F. (1971) The formation and metabolism of 3α, 7α-dihydroxy-5β-cholestan-26-oic acid in man. *J. Clin. Invest.* **50**, 2015–2055
(55) Mitropoulos, K. A. & Myant, N. B. (1967) The formation of lithocholic acid, chenodeoxycholic acid and α- and β-muricholic acids from cholesterol incubated with rat-liver mitochondria. *Biochem. J.* **103**, 472–479
(56) Mitropoulos, K. A., Avery, M. D., Myant, N. B. & Gibbons, G. F. (1972) The formation of cholest-5-ene-3β,26-diol as an intermediate in the conversion of cholesterol into bile acids by liver mitochondria. *Biochem. J.* **130**, 363–371
(57) Makino, I., Sjövall, J., Norman, A. & Strandvik, B. (1971) Excretion of 3β-hydroxy-5-cholanoic acid and 3α-hydroxy-5α-cholanoic acid in urine of infants with biliary atresia. *FEBS Lett.* **15**, 161–164
(58) Sharp, H., Carey, J., Peller, J. & Krivit, W. (1968) Lithocholic acid in meconium. *Pediat. Res.* **2**, 293
(59) Gustafsson, J-A. & Sjövall, J. (1969) Identification of 22-, 24- and 26-hydroxycholesterol in the steroid sulphate fraction of faeces from infants. *Eur. J. Biochem.* **8**, 467–472
(60) Boyd, G. S. & Percy-Robb, I. W. (1971) Enzymatic regulation of bile acid synthesis. *Amer. J. Med.* **51**, 580–587
(61) Percy-Robb, I. W. & Boyd, G. S. (1967) The enterohepatic circulation of bile acids. In *The Liver*; *Colston Papers* No. 19 (Read, A. E., ed.), pp. 11–17, Butterworth and Co., London
(62) Myant, N. & Eder, H. (1961) The effect of biliary drainage upon the synthesis of cholesterol and the liver. *J. Lipid Res.* **2**, 363–368
(63) Danielsson, H., Einarsson, K. & Johansson, G. (1967) Effect of biliary drainage on individual reactions in the conversion of cholesterol to taurocholic acid. Bile acids and steroids 180. *Eur. J. Biochem.* **2**, 44–49

(64) Boyd, G. S., Scholan, N. A. & Mitton, J. R. (1969) Factors influencing cholesterol 7α-hydroxylase activity in the rat liver. *Proc. 3rd Int. Symp. Drugs Affecting Lipid Metab.* (Holmes, W. L., Carlson, L. A. & Paoletti, R., eds.), Plenum Press, New York and London
(65) Heaton, K. W., Austad, W. I. & Lack, L. (1968) Enterohepatic circulation of ^{14}C-labelled bile salts in disorders of the small bowel. *Gastroenterology* **55**, 5–16
(66) Dowling, R. H., Mack, E. & Small, D. M. (1970) Effects of controlled interruption of the enterohepatic circulation of bile salts by biliary diversion and by ileal resection on bile salt secretion, synthesis and pool size in the rhesus monkey. *J. Clin. Invest.* **49**, 232–242
(67) Percy-Robb, I. W., Jalan, K. A., McManus, J. P. A. & Sircus, W. (1971) Effect of ileal resection on bile salt metabolism in patients with ileostomy following proctocolectomy. *Clin. Sci.* **41**, 371–382
(68) Bergström, S. & Danielsson, H. (1958) On the regulation of bile acid formation in the rat liver. *Acta Physiol. Scand.* **43**, 1–7
(69) Shefer, S., Hauser, S., Bekersky, I. & Mosbach, E. H. (1969) Feedback regulation of bile acid synthesis in the rat. *J. Lipid Res.* **10**, 646–655
(70) Frantz, I. D., Jr. & Schroepfer, G. J., Jr. (1967) Sterol biosynthesis. *Annu. Rev. Biochem.* **36**, 691–726
(71) Bucher, N. L. R., McGarrahan, K., Gould, E. & Loud, A. V. (1959) Cholesterol biosynthesis in preparations of liver from normal, fasting, X-irradiated, cholesterol-fed, Triton- or Δ^4-cholesten-3-one-treated rats. *J. Biol. Chem.* **234**, 262–267
(72) Weis, H. J. & Dietschy, J. M. (1969) Failure of bile acids to control hepatic cholesterogenesis. Evidence for endogenous cholesterol feedback. *J. Clin. Invest.* **48**, 2398–2408
(73) Shefer, S., Hauser, S., Lapar, V. & Mosbach, E. H. (1973) Regulatory effects of sterols and bile acids on hepatic 3-hydroxy-3-methylglutaryl CoA reductase and cholesterol 7α-hydroxylase in the rat. *J. Lipid Res.* **14**, 573–580
(74) Boyd, G. S. & Oliver, M. F. (1960) Various effects of thyroxine analogues on the heart and serum cholesterol in the rat. *J. Endocrinol.* **21**, 25–32
(75) Hellström, K. & Lindstedt, S. (1964). Cholic acid turnover and biliary bile-acid composition in humans with abnormal thyroid function. *J. Lab. Clin. Med.* **63**, 666–679
(76) Eriksson, S. (1957) Influence of thyroid activity or excretion of bile acids and cholesterol in the rat. *Proc. Soc. Exp. Biol. Med.* **94**, 588–589
(77) Strand, O. (1962) Influence of propylthiouracil and D- and L-triiodothyronine on excretion of bile acids in bile fistula rats. *Proc. Soc. Exp. Biol. Med.* **109**, 668–672
(78) Björkhem, I., Danielsson, H. & Gustafsson, J. (1973) On the effect of thyroid hormone on 26-hydroxylation of C_{27}-steroids in rat liver. *FEBS Lett.* **31**, 20–22
(79) Rudman, D. & Kendall, F. E. (1957) Bile acid content of human serum. I. Serum bile acids in patients with hepatic disease. *J. Clin. Invest.* **36**, 530–537
(80) Kaplowitz, N., Kok, E. & Javitt, N. B. (1973) Postprandial serum bile acid for the detection of hepatobiliary disease. *J. Amer. Med. Ass.* **225**, 292–293
(81) Carey, J. B., Jr. (1958) The serum trihydroxy-dihydroxy bile acid ratio in liver and biliary tract disease. *J. Clin. Invest.* **37**, 1494–1503
(82) Vlahcevic, Z. R., Miller, J. R., Farrar, J. T. & Swell, L. (1971) Kinetics and pool size of primary bile acids in man. *Gastroenterology* **61**, 85–90
(83) McCormick, W. C., Cooper Bell, Jr., C., Swell, L. & Vlahcevic, Z. R. (1973) Cholic acid synthesis as an index of the severity of liver disease in man. *Gut* **14**, 895–902
(84) Hellström, K. & Lindstedt, S. (1966) Studies on the formation of cholic acid in subjects given standardised diet with butter or corn oil as dietary fat. *Amer. J. Clin. Nutr.* **18**, 46–59
(85) Kottke, B. A. (1969) Differences in bile acid excretion—Primary hypercholesterolemia compared to combined hypercholesterolemia and hypertriglyceridemia. *Circulation* **40**, 13–20
(86) Einarsson, K. & Hellström, K. (1972) The formation of bile acids in patients with three types of hyperlipoproteinaemia. *Eur. J. Clin. Invest.* **2**, 225–230

Essays Med. Biochem. **1**, 81–103
Printed in Great Britain

Hyperlactataemia and Lactic Acidosis*

By H. A. KREBS
Metabolic Research Laboratory, Nuffield Department of Clinical Medicine, Radcliffe Infirmary, Oxford OX2 6HE, U.K.

and H. F. WOODS
M.R.C. Clinical Pharmacology Unit and University Department of Clinical Pharmacology, Radcliffe Infirmary, Oxford OX2 6HE, U.K.

and K. G. M. M. ALBERTI
University Department of Chemical Pathology, Southampton General Hospital, Southampton SO9 4XY, U.K.

Introduction

The quantities of lactic acid formed in the human body can be very high. Severe exercise may produce over 30g of lactic acid within minutes [Hill (1); Åstrand & Rodahl (2)]. Without major exercise the human adult produces probably about 50g of lactic acid in 24h, mainly in the blood cells, though rather higher rates have been computed (18). Yet in the healthy organism the blood lactate concentration varies within narrow limits. When, after severe exercise, the lactate content of the blood has risen more than ten-fold the normal lactate concentration is re-established within less than 90min. There are however pathological states where the regulation of lactate metabolism is disturbed and where the concentration of blood lactate is elevated. This essay is concerned with the occurrence of such raised lactate concentrations in the blood.

The conditions under which blood lactate changes are usefully divided into two groups: hyperlactataemia, characterized by persistent raised blood lactate, usually below 5mM, without changes in blood pH; lactic acidosis, characterized by persistent raised blood lactate, usually above 5mM, with lowered blood pH (3–8).

* This essay is not a comprehensive review. An excellent review published in 1970 is that of Oliva (4). The present aim is not to replace Oliva's article but to give, on the basis of present physiological and biochemical knowledge, a coherent account of the various aspects of the accumulation of lactate in the blood and thus to help the understanding of some clinical situations.

Physiological Aspects

Historical note

The first to establish the occurrence of lactate in mammalian blood were Gaglio in 1886 (9) and Berlinerblau in 1887 (10). Independently from each other these authors isolated lactic acid as zinc lactate and established the correct order of magnitude of about 1 mM-lactate in dog and rabbit blood. Gaglio even noted that the concentration is lower in the blood of starved animals. In those early days, at least 85 ml of blood was necessary (most analyses were done on between 100 and 200 ml) and one analysis took many days to complete. Today, since 1960 (11), thanks to the analytical use of pure enzymes, 0.1 ml of blood, or even less, and a few minutes are sufficient to give rather more accurate values (11).

Before the availability of pure enzymes the method of Clausen (12), especially as modified by Friedemann *et al.* (13), and the method of Mendel & Goldscheider (14), especially as modified by Barker & Summerson (15), facilitated decisive progress in the study of lactic acid metabolism. These methods depend on the conversion of lactic acid into acetaldehyde, and either titrimetric (12, 13) or colorimetric (14, 15) determination of the latter.

Origin of lactate

Small amounts of lactate are contained in food such as meat and milk products but the main source of lactate in the body is the degradation of carbohydrate. The formation of lactate (by glycolysis) is an energy (i.e. ATP)-supplying reaction which is independent of the presence of O_2, but the ATP yield is low. Although the complete oxidation of 1 molecule of glucose supplies up to 38 molecules of ATP, glycolysis supplies only 2 molecules when glucose is the precursor and 3 molecules when glycogen is the precursor. In most animal tissues glycolysis is therefore a minor subsidiary source of ATP under normal conditions. Glycolysis rises in hypoxia when the aerobic generation of ATP is curtailed.

In the absence of oxygen all metabolically active tissues can form lactic acid though the rates vary from tissue to tissue. In the healthy organism hypoxic conditions may occur in striated muscle during exercise when the circulation is insufficient to saturate the tissue with oxygen. Aerobically, i.e. under normal conditions, lactic acid production is low compared with the maximum capacity except in red and white blood cells.

Precise information on the amounts of lactate formed in the human body is lacking. Several authors [Kreisberg (16–18); Searle (19)] have used the isotope-dilution technique to estimate the daily turnover of lactate and came to the conclusion that a human adult weighing 70 kg produces about 140 g of lactic acid in 24 h. Kreisberg computes the production of the human organs as shown in Table 1. The total is still lower than the estimated daily lactic acid production of 140g cited above. The isotope-dilution methods used by Kreisberg (16–18) and Searle (19) in fact measure the sum of lactate plus pyruvate turnover, as pointed out by Searle. The isotope-dilution technique would give reliable results on lactate production if there were no exchange reactions which dilute the labelled substance. Under the conditions of Kreisberg's tests (involving infusion of labelled lactate over a 4h period) such exchanges occur on a very substantial, but not precisely defined scale, and for this reason the 'lactate turnover' is liable to be overestimated, even grossly overestimated, as the following considerations show.

Owing to the high activity of lactate dehydrogenase and the ready reversibility of the reaction catalysed by this enzyme, labelled lactate rapidly exchanges with pyruvate in all metabolically active tissues. Hence labelled lactate is diluted not only with lactate but also with all the pyruvate which arises in the body, and the amounts of pyruvate can be very large. At any one time the pyruvate in the body is approx. 10% of the amount of lactate but the flux through the pyruvate pool is very rapid because the pathways of complete degradation of many amino acids (alanine, serine, glutamate, glutamine, aspartate, arginine, asparagine, proline, threonine, leucine, phenylalanine, tyrosine), and of glucose and other hexoses all pass through pyruvate, as does the synthesis of fatty acids from glucose (Scheme 1). On the other hand gluconeogenesis from some amino acids (glutamate, glutamine, aspartate, asparagine, arginine, proline, threonine, histidine, phenylalanine, tyrosine) does not involve pyruvate whereas that from a few amino acids, shown in Scheme 2, does.

The outcome of these considerations is that the high rate of lactate turnover computed by Kreisberg (18), indicative as it is of the turnover of lactate plus pyruvate, does not supply reliable information on lactate production in the body. Such information is still limited; there is reliable information on lactate formation by whole blood and by various corpuscular elements of blood. As for lactate formation by other tissues this would require measurements of the

Table 1. *Lactic acid production by the major organs of the human body*

The values are those computed by Kreisberg (18).

Organ		Lactic acid (g formed per day)
Blood		33.5
Brain		20.0
Skeletal muscle		18.5
Skin		33.5
Renal medulla		1.7
Intestinal mucosa		8.4
	Sum	**115.6**

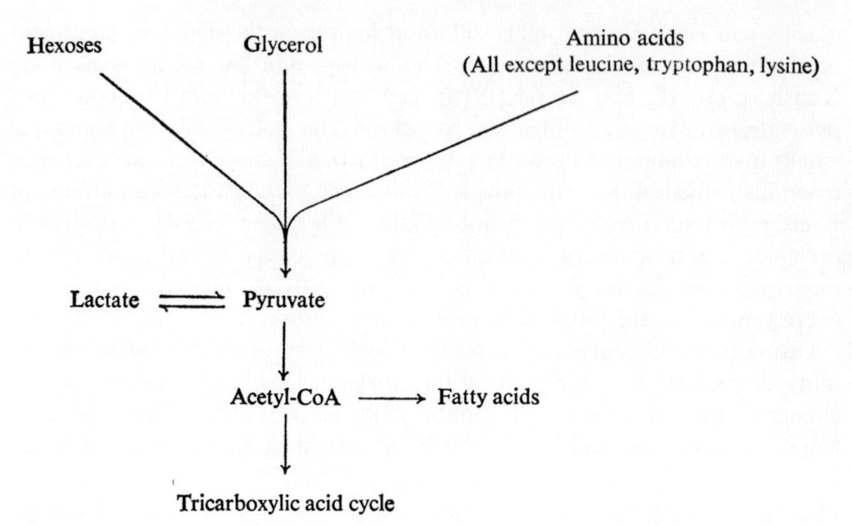

Scheme 1. *Pyruvate as an intermediate in the degradation of fuels of respiration and in fatty acid synthesis*

Owing to the high capacity of lactate dehydrogenase the pyruvate formed as an intermediate from various precursors exchanges almost quantitatively with lactate. Thus the lactate and pyruvate pools are inseparable.

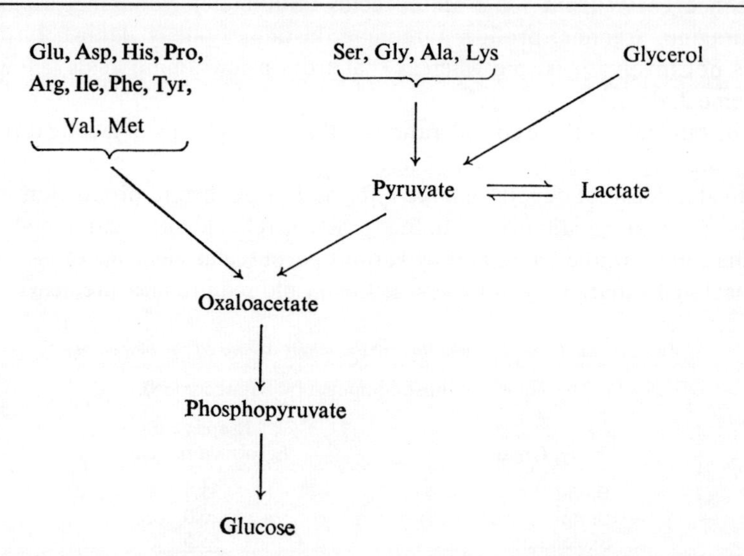

Scheme 2. *Pyruvate as an intermediate in gluconeogenesis from amino acids and glycerol*

The scheme indicates that labelled lactate exchanges with pyruvate arising as an intermediate in gluconeogenesis from certain, but not all, glucogenic precursors. See also legend of Scheme 1.

arteriovenous differences, and the calculations must take into account lactate formation during the passage from the arteries to the veins.

Isolated tissues, such as the perfused liver or the perfused kidney or slices, all produce some lactate if the perfusion or suspension medium is free from lactate but they cease to produce lactate when the medium contains physiological concentrations of lactate. Definite aerobic lactate producers are the exercised muscle, renal medulla, intestinal mucosa, retina and tumour tissue.

Utilization of lactate

There are two main metabolic processes by which lactate is utilized: the synthesis of carbohydrate (gluconeogenesis) and complete combustion. A further pathway, probably a minor one, is the conversion into fatty acids via pyruvate and acetyl-CoA (20–22).

Whether or not lactate is completely oxidized depends on the state of activity of pyruvate dehydrogenase, an enzyme which occurs in an active and a non-active form, the latter being the phosphorylated product of the former. Wieland (23) found the distribution of the two forms in rat tissues to be as shown in Table 2. The enzyme is mainly inactive in liver and kidney, but is active in tissues which usually burn glucose as a fuel of respiration, such as brain and to a lesser extent cardiac muscle. The main factor controlling the state of activity of pyruvate dehydrogenase is the ATP concentration in the tissue. If this is maintained by the combustion of fuels other than carbohydrate then carbohydrate is not burned. Thus the ultimate fate of lactate depends to some extent on the nutritional state of the organism. If the diet is low in carbohydrate, or during starvation, the organism must husband its carbohydrate resources in order to have them available as a source of energy for nervous tissue and blood cells, tissues which cannot survive without carbohydrate in these circumstances. The regulatory mechanisms of gluconeogenesis are so adjusted that any lactate formed in the body is reconverted into carbohydrate at the expense of the energy of oxidation of fat. On the other hand, when the diet contains carbohydrate in abundance carbohydrate is burned.

Table 2. *Activity of pyruvate dehydrogenase in rat tissues* (*Wieland*, 23)

One unit is equal to 1 μmol of acetyl-CoA formed/min per mg of DNA at 37°C.

	Well-fed		Starved	
	Units	% activity	Units	% activity
Heart muscle	4.9	70	3.8	14
Kidney	0.7	68	0.8	13
Liver	0.9	17	0.6	10
Adipose tissue	1.2	20	1.2	7
Brain	1.6	68	1.6	70

Liver and kidney cortex are the only tissues which are capable of synthesizing glucose from lactate. On a weight-to-weight basis the gluconeogenic capacity of liver and kidney cortex are of the same order of magnitude, but because of the much larger quantity of liver tissue the liver is evidently the main gluconeogenic organ. Nevertheless, there are situations where renal gluconeogenesis may be more important than the relatively small amount of renal gluconeogenic tissue suggests. The blood supplies to the kidneys and to the liver are about equal (about 25% of the resting cardiac output for each tissue), and this means that the amounts of lactate offered to the two tissues are about equal. Since the rates of metabolic reactions depend not only on the capacity of the enzymes but also on the amounts of substrate available, the kidneys may make special contributions to gluconeogenesis, especially under conditions of severe exercise. This assumption is supported by the fact that the renal capacity for gluconeogenesis adaptively increases on severe exercise (24, 25).

Attempts have been made *in vivo* with the help of ^{14}C-labelled lactate to decide how much lactate is oxidized and how much is converted into glucose under various circumstances (16–18). However, owing to complex exchange reactions [as discussed by Weinman *et al.* (26), Krebs *et al.* (27) and Steele (28)] such tests do not give satisfactory results. They are liable to overestimate grossly the conversion of lactate into carbon dioxide and to underestimate grossly glucose formation.

Regulation of blood lactate concentration

Whereas the formation of lactate is essentially an energy-supplying reaction the removal of lactate is largely a salvage operation, i.e. the recovery of a substance not immediately needed as a fuel, to be stored as glycogen and to be made available as glucose to the organs when required. The capacity of this salvage process can be calculated from the rate at which lactate which has accumulated after severe exercise disappears from the body fluids. The value calculated is of the order of 330 g of lactic acid per 24 h for an adult of 70 kg (1, 29), which is very much higher than lactate production under normal conditions. This capacity for disposing of accumulated lactate is a major factor limiting severe exercise.

Thus the regulation of the lactate concentration in the blood and tissues is mainly a matter of lactate removal. Removal from the blood is never complete. It stops when the concentration in the blood is about 1 mM in the well-fed state and somewhat lower in the fasted state. The reasons for these limits are not yet properly understood. The reactions that limit the rate of lactate disposal are, as already discussed, the pyruvate-removing enzymes, i.e. pyruvate dehydrogenase and, in some tissues, pyruvate carboxylase. Both depend on the pyruvate concentration, and as the concentrations of lactate and pyruvate are interdependent and are present in approximately constant proportions (of about 1:10 in favour of lactate) a low pyruvate concentration in the tissues may prevent the complete removal of lactate. Pyruvate is one of the activators of pyruvate dehydrogenase (23). At pyruvate concentrations below 0.1 mM the dehydrogenase becomes largely inactive because of the conversion of the active into the inactive form. Factors causing inhibition of the residual active form are the rises in the tissue concentrations of acetyl-CoA (30) and long-chain fatty acid derivatives (31), which occur when fat largely replaces carbohydrate as a source of energy, e.g. during starvation.

Normal blood lactate concentrations

At rest the variations in blood lactate concentration are slight [for review see Oliva (4)]. In the normal well-fed human adult the lactate concentration in venous blood is about 1 mM (range 0.6–1.2 mM) (A. S. Rowe, A. Dornhorst & K. G. M. M. Alberti, unpublished work). It is slightly lower after fasting (about 0.7 mM, range 0.4–1.0 mM) (A. S. Rowe, A. Dornhorst & K. G. M. M. Alberti, unpublished work). The pyruvate concentration is usually about one-tenth of that of lactate, except after severe exercise when the [lactate]/[pyruvate] ratio rises.

During an 'all-out' acute physical exercise, such as sprinting or walking on a treadmill with a heavy load, exhaustion, as indicated by an absolute need for rest, usually occurs within 2–5 min. In such a situation the blood lactate concentration is about 10 mM immediately after the exercise. It falls rather rapidly on rest in the first instance because of diffusion into body water. When this diffusion has been completed within 10 or 20 min, the removal is caused by metabolic processes and is relatively slow. Exhaustion is not due to the accumulation of lactate in the blood, but to a combination of factors such as acidification of the tissues, lack of oxygen, and in the last resort the depletion of muscle ATP. Much higher blood lactate concentrations can be reached by some individuals. Turrell & Robinson (32) report that concentrations of up to 22 mM are not uncommon in trained athletes after severe exercise.

Regulation of the [lactate]/[pyruvate] ratio

The [lactate]/[pyruvate] ratio reflects the redox state of the cytoplasmic NAD couple, defined as the ratio

$$\frac{[\text{free NAD}^+]}{[\text{free NADH}]}$$

The value of this ratio is of the same order in most tissues, between 300 and 1000 at pH 7.0. 'Free' is as opposed to protein-bound NAD (33). Owing to the high activity of lactate dehydrogenase in animal tissues the reversible reaction catalysed by this enzyme

$$\text{Lactate} + \text{NAD}^+ \rightleftharpoons \text{pyruvate} + \text{NADH} + \text{H}^+ \qquad (1)$$

is usually at near-equilibrium. This implies that the redox state of the NAD couple can be calculated from the relation

$$\frac{[\text{Lactate}]}{[\text{Pyruvate}]} = \frac{1}{K_{\text{LDH}}} \times \frac{[\text{NADH}][\text{H}^+]}{[\text{NAD}^+]} \tag{2}$$

where K_{LDH} is the equilibrium constant of reaction (1).

The relative constancy of the redox state indicates that it is controlled by special mechanisms. The evidence supports the concept that it is related to, and probably mainly regulated by, the phosphorylation state of the cytoplasmic adenine nucleotide system, i.e. the value of the ratio [ATP]/[ADP] [P_i] (34, 35). The regulating link between the phosphorylation state of the adenine nucleotide system and the redox state of the NAD couple is provided by the glyceraldehyde phosphate dehydrogenase and 3-phosphoglycerate kinase reactions, catalysing the overall reaction

$$\begin{aligned}&\text{Glyceraldehyde phosphate} + \text{ADP} + \text{P}_i + \text{NAD}^+ \\ &\quad \rightarrow \text{3-phosphoglycerate} + \text{ATP} + \text{NADH} + \text{H}^+\end{aligned} \tag{3}$$

The capacities of the two enzymes catalysing reaction (3) are high and the reactants of the system are therefore at near-equilibrium *in vivo*. At equilibrium the following relation holds:

$$\frac{[\text{NAD}^+]}{[\text{NADH}][\text{H}^+]} = \frac{1}{K} \times \frac{[\text{3-Phosphoglycerate}]}{[\text{Glyceraldehyde phosphate}]} \times \frac{[\text{ATP}]}{[\text{ADP}][\text{P}_i]} \tag{4}$$

where K is the product of the equilibrium constants of the glyceraldehyde phosphate dehydrogenase system and the 3-phosphoglycerate kinase system. Combination of eqns. (2) and (4) gives

$$\frac{[\text{Lactate}]}{[\text{Pyruvate}]} = \text{constant} \times \frac{[\text{Glyceraldehyde phosphate}]}{[\text{3-Phosphoglycerate}]} \times \frac{[\text{ADP}][\text{P}_i]}{[\text{ATP}]} \tag{5}$$

Thus on the assumption that the [lactate]/[pyruvate] ratio in blood corresponds to that in tissues, the phosphorylation state of the adenine nucleotides plays a key role in the control of the ratio. The phosphorylation state in turn is regulated by the mitochondrial respiratory chain. In accordance with this concept, the high [lactate]/[pyruvate] ratios in hypoxia correlate with low values of the phosphorylation state. Incidentally, it follows from eqn. (2) that, other factors being equal, the rise in [H^+] also raises the [lactate]/[pyruvate] ratio.

Comment on the relations between H^+ ion formation and acidosis

Enormous quantities of H^+ ions are produced continuously in living cells, but this normally does not lead to acidosis because production of H^+ ions is balanced by removal. Indeed the turnover of H^+ ions is greater than that of any other intermediary metabolite. This is so because the most frequent single reaction in aerobic cells is the conversion of ATP into ADP and P_i which at the pH of the cell is accompanied by a stoicheiometric formation of H^+ ions according to the equation

$$ATP^{4-} + H_2O \rightarrow ADP^{3-} + HPO_4^{2-} + H^+ \qquad (6)$$

As some phosphate arises in the form of $H_2PO_4^-$, a fraction of the ATP hydrolysed, possibly 30%, is not accompanied by the production of H^+ ions but reacts according to the equation

$$ATP^{4-} + H_2O \rightarrow ADP^{3-} + H_2PO_4^- \qquad (7)$$

Large additional amounts of H^+ are formed in the respiratory chain when Fe^{3+} reacts with a flavoprotein or other hydrogen donors according to the equation

$$2\,Fe^{3+} + 2\,H \rightarrow 2\,Fe^{2+} + 2\,H^+ \qquad (8)$$

Further, in all dehydrogenase reactions in which NAD or NADP is the hydrogen acceptor H^+ ions are formed according to the equation

$$NAD^+ \text{ (or } NADP^+) + 2\,H \rightarrow NADH \text{ (or } NADPH) + H^+ \qquad (9)$$

Considering that for 1 molecule of oxygen used, up to 6 molecules of ATP are formed of which 4 react according to eqn. (6), and that 2 molecules of NAD or NADP as well as 4 molecules of Fe^{3+} are involved in cell metabolism, at least 10 H^+ ions are formed per molecule of O_2. These are removed when ATP is resynthesized and when nicotinamide nucleotides and reduced cytochromes are reoxidized, so that the pH remains virtually constant.

The order of magnitude of the turnover of H^+ ions can thus be assessed from the oxygen consumption, and in the human adult with a caloric expenditure of 9200 kJ/day must be of the order of 150 g per day.

It is of interest that the turnover of ATP at a daily consumption of 9200 kJ, on the assumption that 6 ATP molecules are formed for each O_2 molecule used, is 120 mol of ATP per day or 61.7 kg of disodium ATP per day!

In addition to the turnover of H^+ ions in the course of dehydrogenations and oxidative phosphorylation, there is a turnover associated with non-oxidative reactions such as the formation of lactate from carbohydrate. This is obligatorily coupled with the synthesis of ATP according to the equations

$$\text{Glucose} + 2\,HPO_4^{2-} + 2\,ADP^{3-} \rightarrow 2\,\text{lactate}^- + 2\,ATP^{4-} \qquad (10)$$

$$\text{Glucose} + 2\,H_2PO_4^- + 2\,ADP^{3-} \rightarrow 2\,\text{lactate}^- + 2\,ATP^{4-} + 2\,H^+ \qquad (11)$$

Reaction (10) is always followed by reaction (6) and reaction (11) by reaction (7) and it follows that most of the H^+ ions are not formed during glycolysis but by reaction (6). All the reactions discussed occur within the cell, and neutralization is therefore brought about by intracellular buffers, in the first instance bicarbonate. In muscle, carnosine and anserine can also be of importance. Any bicarbonate used is replaced by plasma bicarbonate which readily enters the cells.

Lactate production and ATP synthesis, although stoicheiometric as formulated in eqns. (10) and (11), are not stoicheiometric over long periods because reactions (10) and (11) occur at the same time as reactions (6) and (7), which means that the same molecules of ADP and P_i are used over and over again.

In the healthy organism lactate formation is always followed by lactate removal, either by oxidation:

$$\text{Lactate}^- + H^+ + 3\ O_2 \rightarrow 3\ CO_2 + H_2O$$

or by glucose synthesis:

$$2\ \text{Lactate}^- + 2\ H^+ \rightarrow \text{glucose}$$

and thus the H^+ ions formed together with lactate disappear when the lactate is removed so that there is no change in pH.

Reactions (10) and (11) are simplified accounts of the actual events because they describe only the overall balance of glycolysis. There are intermediary steps where H^+ ions are formed, e.g. the hexokinase, the phosphofructokinase and (twice) the glyceraldehyde phosphate dehydrogenase reactions. There are other steps where equivalent amounts of H^+ ions are consumed, e.g. in the pyruvate kinase reaction (twice) and lactate dehydrogenase reaction (twice). (For detailed formulation of the intermediary reactions see refs. 36 and 37.)

Compared with the amounts of H^+ ions formed in the body (more than 150 g per day, equivalent to 13.5 kg of lactic acid) the amounts of lactic acid necessary to produce acidosis are small. Acidosis occurs when an excess of anions such as β-hydroxybutyrate, lactate, sulphate or chloride must be excreted with the urine and when there are not sufficient cations to neutralize them. The intermediary formation of H^+ ions on a very large scale does not lead to acidosis because under normal conditions it is automatically coupled with the removal of H^+ ions and there is therefore no loss of anions by urinary excretion.

Practical Aspects

Stasis produced by a tourniquet, by muscular contraction in the limb during blood collection, or by storage of the collected blood, all cause the lactate concentration to rise (38–41). Blood collection for lactate determination must therefore be carried out in such a way that the above factors do not significantly affect the value *in vivo*, and deproteinization by $HClO_4$ should be carried out immediately after sampling.

Arterial blood contains slightly less lactate than venous blood collected from the forearm, probably because of lactate production by the red and white cells during the passage through the capillaries, though production by the tissues, especially by the skin, cannot be excluded. Human blood at body temperature forms approx. 2 μmol of lactate/ml per h. Whole human blood contains very slightly more lactate than plasma because there is a slight concentration gradient from the site of the formation of lactate within the cells to the plasma. In general the lactate space equals the total body water space and this implies that the gradient between tissues and plasma is slight under normal conditions. It can, of course, become very substantial after severe exercise or in stasis.

Pathological Aspects

General considerations

The maintenance of normal blood lactate concentrations largely depends on the intactness of the processes that dispose of the lactate continuously formed in the body at a daily rate of 30–50 g in the resting human adult and at greatly increased rates on severe exercise. As discussed above the lactate-disposing processes are essentially of two kinds: complete oxidation (regulated at the stage of pyruvate dehydrogenase) and resynthesis to carbohydrate (regulated by the blockage of pyruvate dehydrogenase in liver and kidney and by the direction of the pyruvate to carbohydrate) via the pyruvate carboxylase reaction, as illustrated by Scheme 3. The synthesis of fatty acids indicated in Scheme 3 is probably a very minor process. How large a fraction of pyruvate enters each of the two pathways must depend on the activities of the two key enzymes and these in turn depend very much on the nutritional state. According to Table 2 the state of activity of pyruvate dehydrogenase decreases in some tissues during starvation. Gluconeogenesis on the other hand is known to increase during starvation. The capacity of the body to dispose of lactate by oxidation is limited by its energy requirements. Hence after exercise much of the lactate that has accumulated must be resynthesized to carbohydrate through the 'Cori-cycle', i.e. in the liver, rather than burned, because the fuel requirements are too low. For this reason the role of the liver in lactate metabolism is a matter of special importance in the present context; on account of its high gluconeogenic capacity, it is by far the most important site of lactate disposal.

There is no precise information on the maximum gluconeogenic capacity of the human liver. Rat liver can convert up to 4 μmol of lactate into glucose/min per g or 3.6 g of lactate/24 h per 200 g (42). Human liver, like that of other

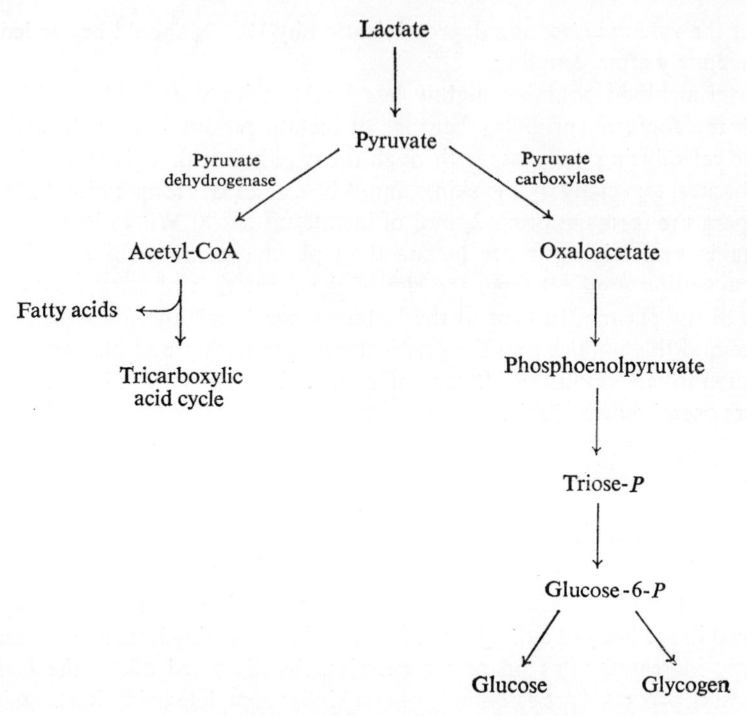

Scheme 3. *Main pathways of lactate metabolism*

larger animals, has generally lower metabolic rates per unit of weight and the proportion of liver to whole body weight is smaller in man than in small animals. But taking these factors into consideration the gluconeogenic capacity of the human liver would still be expected to be up to at least 240 g of lactate daily.

Two relevant consequences arise from this role of the liver. First, as the liver is the main site of lactate disposal it is to be expected that the blood lactate concentration is raised in liver diseases. Secondly, as the capacity for disposing of lactate is normally adjusted to conditions of severe exercise and is therefore a multiple of the requirements at rest, liver diseases that eliminate a major proportion of liver function usually leave sufficient residual capacity to cope with excess of lactate in the non-stressed state. If minor increases in blood lactate concentrations occur in a variety of pathological states they are likely to be a matter of modified regulation of the blood lactate concentration rather than the inability of the liver to convert lactate into carbohydrate.

A third point of relevance to lactataemia is the fact that the liver has a high potential capacity for producing lactate and instead of disposing of lactate it can become a lactate-producing organ. In the normal liver in the well-fed state, when glycogen is being stored, this capacity is not manifest. The high capacity to produce lactic acid is connected with one of the major liver functions, the conversion of carbohydrate into fat. This process consists of two phases (Scheme 4). In the first series of reactions glycogen is converted into acetyl-CoA through the stages of glycolysis plus the pyruvate dehydrogenase reaction. In the second phase acetyl-CoA is converted into fatty acids. The first phase is a degradative process yielding energy in the form of ATP. The second is a biosynthetic process requiring ATP (plus reducing equivalents in the form of NADPH). When the oxygenation of the liver is inadequate, the second phase becomes defective and pyruvate, instead of undergoing oxidation to acetyl-CoA, is reduced to lactate.

Experimentally the separation of glycogen degradation to lactate from fatty acid synthesis is strikingly achieved by depriving the rat liver of oxygen, or of the means of utilizing oxygen, e.g. by poisoning it with hydrocyanic acid (43). Then the liver, instead of being a lactate-consuming organ, becomes a lactate-producing organ.

It should be emphasized that the liver forms lactic acid only when there is a store of glycogen. Glucose is hardly used by the liver, in contrast with fructose. The latter is the only monosaccharide that is readily metabolized because the hexokinase activity of the liver is low, whereas the special fructokinase which converts fructose into fructose 1-phosphate is highly active.

Thus a major cause of severe lactic acidosis can be the transformation of the liver from a lactate-consuming into a lactate-producing organ (44). There are, of course, intermediate stages where the lactate-consuming function may not be abolished but seriously diminished.

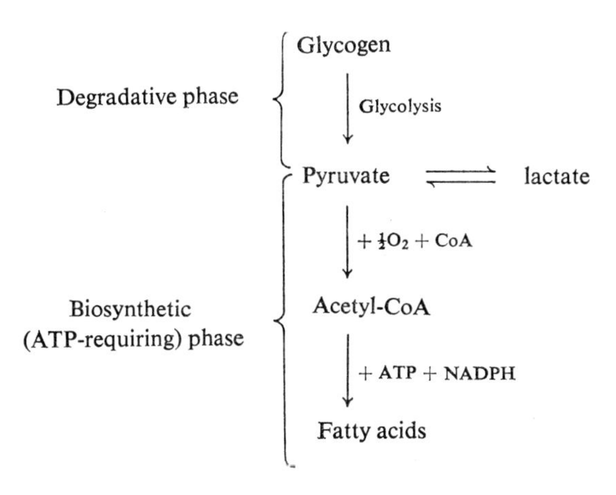

Scheme 4. *Pathway of hepatic conversion of carbohydrate into fatty acids*

Diseases of the liver

In the light of the preceding section it is not surprising that the most severe disruption of lactate metabolism, with pronounced acidosis occurs as a consequence of hypoxia, appearing in congestive heart failure when oxygenation of the blood is inadequate and when the circulation of the liver is hindered by hepatic oedema. This situation affects every hepatocyte, in contrast with many other liver diseases, for example liver cirrhosis where a number of fairly competent hepatocytes persist even in advanced states of the disease. Many of the cases of severe lactate acidosis described in the literature belong to this category of cardiovascular insufficiency. Lactate concentrations in the blood may exceed 25 mM and blood pH falls to below 7.0.

The clinical picture is dominated by the disorder which gives rise to the severe disturbance of hepatic metabolism, upon which are superimposed in severe cases the characteristic signs of metabolic acidosis, hyperpnoea and tachypnoea. The first cases of lactic acidosis were described by Huckabee (45) in 1961. In 1964 Tranquada (5, 6) described and analysed a series of 47 cases of non-ketotic acidosis. The first laboratory finding to point to the underlying metabolic abnormality was an unexplained discrepancy between the plasma chloride and the plasma cation concentrations. The search for the missing anion identified it as lactate. The rather late discovery of lactic acidosis is no doubt connected with the fact that lactate determinations were not one of the routine tests in a clinical laboratory. Once clinicians become aware of the possibility of lactic acidosis cases are regularly discovered, for example at the Radcliffe Infirmary, Oxford, they are diagnosed at the rate of about 2 per month and apparently more frequently in other areas, especially those where chronic alcoholism is a major cause of liver disease.

Blood lactate concentrations are raised in a wide spectrum of liver diseases (46–50) but the increase above the normal value of 0.6–1.2 mM is usually moderate, rarely exceeding 2 mM-lactate. A most common cause is alcoholic liver disease. Such patients may also show an impaired clearance of an exogenous lactate load (51, 52).

The assumption of liver damage is hardly sufficient to explain a rise in blood lactate of such patients at rest because the capacity of the liver to remove the amounts of lactate formed at rest should be adequate. It therefore seems more likely that the raised values are caused by a disturbance of regulatory mechanisms. On the basis of the comments under 'Regulation of blood lactate concentration' these are primarily the mechanisms which maintain the normal pyruvate concentration.

Hyperventilation

In man spontaneous hyperventilation causes a rise in blood pH to about 7.70, and this alkalosis is accompanied by an increase in lactate concentration in venous blood by a factor of 2 or 3 (53–57). In experiments on anaesthetized dogs Berry & Scheuer (58) observed increases up to 10 mM, and this was associated with shedding of lactate by the splanchnic bed (i.e. the liver), the concentrations of lactate after 30–45 min being 6.4 mM in the aorta, 5.9 mM in the vena cava and 9.6 mM in the hepatic vein. Blood pressure, hepatic blood flow and hepatic venous oxygen tension fell, indicating that the hyperlactataemia during hyperventilation like that of liver disease is caused by a disturbance of the hepatic circulation, converting the liver from a lactate-removing into a lactate-producing organ.

Shock

Cournand *et al.* (59) reported in 1943 that shock of various types causes a metabolic acidosis with high lactate concentrations in the blood. This was confirmed by many subsequent investigators (60–65). According to Peretz *et al.* (65) the lactate concentration in the blood correlates with the survival rates of the patients. The mortality rate was 100% when the lactate concentration in the blood exceeded 13 mM. It was 90% when the blood lactate concentration was between 9 and 13 mM and 75% when it was between 4.5 and 9 mM. Presumably general tissue hypoxia caused by inadequate blood supply is responsible for excessive lactate production by various tissues and a failure of the liver to remove lactate.

Diabetes

Except in severe ketoacidosis the blood lactate concentration is not much raised in diabetes. In severe ketoacidosis Alberti & Hockaday (66) found values of over 5 mM in 14% of the cases and Watkins *et al.* (67) in 9%. As a rule lactic acid does not make a significant contribution to the development of metabolic acidosis in diabetics because the amounts of lactate excreted in the urine are relatively small, the renal threshold of lactate in man being 6.6 mM [Pitts (68)], and it should be borne in mind that the development of acidosis depends on the excretion of fixed base (i.e. Na^+ and K^+) rather than the formation of intermediary metabolites. Minor increases in blood lactate occur especially in juvenile-onset diabetes (69).

Iatrogenic disturbances

A number of therapeutic agents can occasionally give rise to lactataemia and lactic acidosis. The most common drugs among these are the hypoglycaemic biguanides, such as phenformin. In therapeutic doses phenformin causes mild hyperlactataemia in normal and diabetic subjects, the rise being dose-related (70–73). In the isolated perfused rat liver phenformin inhibits gluconeogenesis from lactate but only at concentrations which are more than 100 times higher than those attained in the blood at therapeutic doses (73).

Considering the large number of diabetics receiving biguanide treatment the occurrence of lactic acidosis is infrequent. This implies that factors other than drug toxicity must be responsible, such as genetic differences, impairment of liver function or impairment of renal excretion of phenformin.

Of special interest among the iatrogenic forms of lactic acidosis are those caused by parenteral nutrition with solutions containing fructose (74). Fructose rather than glucose has been used as a parenteral nutrient because the utilization of fructose does not require insulin. However, the large amounts of fructose reaching the liver can produce a damaging 'side effect' (75–77). The first stage of fructose utilization by the liver is the interaction with ATP, catalysed by fructokinase:

$$\text{Fructose} + \text{ATP} \rightarrow \text{fructose 1-phosphate} + \text{ADP}$$

Fructose 1-phosphate accumulates in the liver and thereby sequesters phosphate. Moreover, the fructokinase reaction can be so fast that it depletes the liver of ATP and phosphate. As ATP and phosphate are essential for maintaining the stability of ADP and AMP, the latter two substances are lost by conversion to inosine monophosphate and adenosine, and in rat liver the total adenine nucleotide content falls from the normal range of 3–4 mM to 1 mM. This loss of ATP, like that caused by hypoxia, converts the liver into an organ which produces lactate from glycogen and from the added fructose.

As fructose has no essential advantage over glucose as a parenteral nutrient (but involves a dangerous side effect) its inclusion in nutrition solutions should be abandoned (74).

Sorbitol and xylitol have also been used as parenteral nutrients and act in principle in the same way as fructose, in that they can cause depletion of adenine nucleotides in the liver (78, 79). Their parenteral use is therefore also to be deprecated.

Inborn errors of metabolism

Abnormalities in lactate metabolism occur in several inborn errors. In glycogen-storage disease type I (glucose 6-phosphatase deficiency) it is connected with the fact that the formation of glucose from glycogen is restricted (if not completely blocked) and thus glycogen can be broken down only to pyruvate and lactate (80–83). Another form is characterized by recurring attacks of severe lactic acidosis in infancy. The cause of this disturbance is obscure. The [lactate]/[pyruvate] ratio is normal but the alanine concentration in the blood is raised (84, 85). A third type (86), noted in adults, shows a permanent elevation of the blood lactate which rises abnormally after exercise and alcoholic ingestion, and the acidosis can be sufficiently severe to lead to coma.

An inborn error of special interest is a case recently described (87) where the cause of lactic acidosis can be established. The patient had intermittent lactic acidosis with hypoglycaemia and a raised pyruvate concentration of 1 mM against the normal value of 0.1 mM. Pyruvate carboxylase in a biopsy sample of the liver had an abnormal K_m value of 3.7 mM, whereas normal human liver was found to have two components of pyruvate carboxylase, one with a K_m of 0.4 mM and a second with a K_m of 3.7 mM. The [lactate]/[pyruvate] ratio was normal. In this case obviously the primary abnormality was in the pyruvate metabolism.

Other forms (88–94) cannot be discussed here. The general inference is that the possibility of lactic acidosis must be borne in mind whenever infants are suspected of an inborn error because of failure to thrive or other unexplained abnormalities.

Other conditions

There are further conditions where the lactate concentration in the blood can be raised (see ref. 4). They include methyl and ethyl alcohol intake, renal diseases, anaemias, leukaemia, eclampsia and thiamine deficiency. In thiamine deficiency the primary abnormality is a raised blood pyruvate concentration (95–97) arising from the requirement of thiamine pyrophosphate as cofactor of pyruvate dehydrogenase, one of the two major enzymes responsible for pyruvate removal. The [lactate]/[pyruvate] ratio is normal in thiamine deficiency (97) which tallies with the concept that this ratio depends on the redox state of the NAD couple, and that the rise in the lactate concentration is secondary in this case to the rise in the pyruvate concentration.

Treatment

The mortality rate in patients with fully developed lactic acidosis is high because of the severity of the underlying causes, circulatory failure, diabetes, liver disease, renal disease. Treatment of the cause has therefore first priority. The second is the correction of pH by intravenous infusion of $NaHCO_3$. Large amounts may be required involving loading the body with sodium and water. To avoid this load the use of the buffer THAM [tris(hydroxymethyl)-aminomethane] has been recommended (4), but whether THAM buffer is more beneficial than bicarbonate is still an open question.

Attempts have been made (4) to accelerate lactic acid removal by intravenous infusion of Methylene Blue which promotes the conversion of lactate into pyruvate, and by haemodialysis or peritoneal dialysis. The latter two procedures alone cannot correct pH rapidly and must therefore be used together with $NaHCO_3$ infusion. They have the advantage of avoiding overloading with sodium and water, but again there is insufficient experience to assess their merits.

The prognosis is better when drugs like phenformin are the causative agents, if the drug is withdrawn early enough. For a fuller discussion of the problem of treatment the reader is referred to the book of Cohen & Woods (99).

Envoy

The minor forms of lactataemia which occur in a variety of diseases are of no major practical clinical importance because there are subsidiary signs of various syndromes. They are neither of diagnostic, prognostic nor of therapeutic relevance.

Major forms of lactataemia and lactic acidosis such as occur in cardiovascular insufficiency, hyperventilation, shock, drug intoxication and some inborn errors are of practical interest because their recognition and treatment may influence the course of the disease. Though lactic acidosis is not very frequent it probably occurs much more often than is at present realized. To diagnose it a method of lactate determination in the blood should be available as a routine and be made whenever a lowered blood pH is suspected.

The analysis of the cause of lactic acidosis brings home the fact that the liver plays a key role in the control of lactate metabolism. The liver is the primary site of every major, and probably most of the minor, disturbances of lactate metabolism. As Berry (44) first clearly recognized, it is the change from a lactate-removing organ to a lactate-producing organ which is at the root of the major disorders of lactate metabolism. It is now understood how this change comes about: inadequate oxygenation (this means inadequate ATP supply) not only prevents gluconeogenesis from lactate but also initiates glycogen degradation to the stage of pyruvate, normally a step in fatty acid synthesis. But in hypoxic conditions pyruvate is reduced to lactate instead of forming acetyl-CoA and fatty acids. A point of practical interest is the fact that lactate production by the liver depends on the glycogen stores of the organ. Hence a low carbohydrate diet is expected to decrease lactate production.

The question has often been raised whether lactic acidosis is a matter of overproduction or underutilization of lactate (98). However, this is not an either–or question. Normally lactate production is compensated by lactate utilization. Even when excessive muscular exercise produces very large amounts of lactic acid within a short time other organs, especially the liver, rapidly remove the extra lactate. The diseased liver may more or less lose the capacity of utilizing lactate, and instead produces it. Thus an element of both underutilization and overproduction contributes to lactic acidosis. Lactataemia without acidosis is usually not a matter of imbalance between production and removal but an abnormality of the regulation of the lactate concentration, the primary event being either a raised pyruvate concentration (as in thiamine deficiency and certain inborn errors of metabolism) or a change in the redox state of the NAD couple (as in alcohol intoxication).

However, imbalance between production and removal is not the only matter of importance in the development of lactic acidosis. A crucial factor, as in ketoacidosis, is the loss of base in the urine. The renal threshold for lactate in man is about 6.6 mM compared with 2 mM for β-hydroxybutyrate (68). Thus base is lost only when the concentration of lactate in blood exceeds 6 mM (unless kidney function is impaired). This accounts for the fact that lactic acidosis as a rule does not develop when the lactate concentration is below 5 mM and that lactataemia up to 5 mM is not accompanied by acidosis.

References

(1) Hill, A. V. (1926) The recovery process after muscular exercise in man. *Harvey Lect.* **20**, 60–77
(2) Åstrand, P-O. & Rodahl, K. (1970) *Textbook of Work Physiology* especially pp. 294–318, McGraw–Hill Book Co.,
(3) Huckabee, W. E. (1956) Control of concentration gradients of pyruvate and lactate across cell membranes in blood. *J. Appl. Physiol.* **9**, 163–170
(4) Oliva, P. B. (1970) Lactic acidosis. *Amer. J. Med.* **48**, 209–225
(5) Tranquada, R. E. (1964) Lactic acidosis. *Calif. Med.* **101**, 450–461
(6) Tranquada, R. E., Grant, W. J. & Peterson, C. R. (1966) Lactic acidosis. *Arch. Int. Med.* **117**, 192–202
(7) Peretz, D. I., McGregor, M. & Dossetor, J. B. (1964) Lactic acidosis: a clinically significant aspect of shock. *Can. Med. Ass. J.* **90**, 673–675
(8) Leading article (1970) *Brit. Med. J.* **4**, 528
(9) Gaglio, G. (1886) Die Milchsäure des Blutes und ihre Ursprungsstätten. *Arch. Anat. Physiol. Physiol. Abt.* **10**, 400–414
(10) Berlinerblau, M. (1887) Über das Verkommen der Milchsäure im Blute und ihre Entstehung im Organismus. *Arch. Exp. Pathol. Pharmakol.* **23**, 333–346
(11) Hohorst, H. J. (1960) Ph.D. Dissertation, University of Marburg
(12) Clausen, S. W. (1922) A method for the determination of small amounts of lactic acid. *J. Biol. Chem.* **52**, 263–280
(13) Friedemann, T. E., Cotonio, M. & Shaffer, P. A. (1927) Determination of lactic acid. *J. Biol. Chem.* **93**, 335–358
(14) Mendel, B. & Goldscheider, I. (1925) Eine kolorimetrische Mikromethode zur quantitativen Bestimmung der Milchsäure im Blut. *Biochem. Z.* **164**, 163–174
(15) Barker, S. B. & Summerson, W. H. (1941) Colorimetric determination of lactic acid in biological material. *J. Biol. Chem.* **138**, 535–554
(16) Kreisberg, R. A., Pennington, L. F. & Boshell, B. R. (1970) Lactate turnover and gluconeogenesis in normal and obese humans. *Diabetes* **19**, 53–63
(17) Kreisberg, R. A., Pennington, L. F. & Boshell, B. R. (1970) Lactate turnover and gluconeogenesis in obesity. *Diabetes* **19**, 64–69
(18) Kreisberg, R. A. (1972) Glucose–lactate interrelations in man. *N. Engl. J. Med.* **287**, 132–137
(19) Searle, G. L. & Cavalieri, R. R. (1972) Determination of lactate kinetics in the human analysis of data from single injection vs. continuous infusion methods. *Proc. Soc. Exp. Biol. Med.* **139**, 1002–1006

(20) Exton, J. H., Corbin, J. G. & Harper, S. C. (1972) Control of gluconeogenesis in liver. V. Effects of fasting, diabetes and glucagon on lactate and endogenous metabolism in the perfused rat liver. *J. Biol. Chem.* **247**, 4996–5003
(21) Clark, D. G., Rognstad, R. & Katz, R. (1974) Lipogenesis in rat hepatocytes. *J. Biol. Chem.* **249**, 2028–2036
(22) Katz, J. & Wals, P. A. (1974) Lipogenesis from lactate in rat adipose tissue. *Biochim. Biophys. Acta* **348**, 344–356
(23) Wieland, O. H., Siess, E. A., Weiss, L., Löffler, G., Patzelt, C., Portenhauser, R., Hartmann, U. & Schirmann, A. (1973) Regulation of the mammalian pyruvate dehydrogenase complex by covalent modification. *Symp. Soc. Exp. Biol.* **27**, 371–400
(24) Krebs, H. A. (1963) Renal gluconeogenesis. *Advan. Enyzme Regul.* **1**, 385–400
(25) Krebs, H. A. & Yoshida, T. (1963) Muscular exercise and gluconeogenesis. *Biochem. Z.* **338**, 241–244
(26) Weinman, E. O., Strisower, E. H. & Chaikoff, I. L. (1957) Conversion of fatty acids to carbohydrate: application of isotopes to this problem and role of the Krebs cycle as a synthetic pathway. *Physiol. Rev.* **37**, 252–272
(27) Krebs, H. A., Hems, R., Weidemann, M. J. & Speake, R. N. (1966) The fate of isotopic carbon in kidney cortex synthesizing glucose from lactate. *Biochem. J.* **101**, 242–249
(28) Steele, R. (1971) *Tracer Probes in Steady State Systems*, Charles C. Thomas, Springfield, Ill.
(29) Peters, J. P. & Van Slyke, D. D. (1946) *Quantitative Clinical Chemistry Interpretations*, 2nd edn., vol. 1, Baillière, Tindall and Cox, London
(30) Garland, P. B. & Randle, P. J. (1964) Control of pyruvate dehydrogenase in the perfused rat heart by the intracellular concentration of acetyl-coenzyme A. *Biochem. J.* **91**, 6C
(31) Portenhauser, R. & Wieland, O. (1972) Regulation of pyruvate dehydrogenase in mitochondria of rat liver. *Eur. J. Biochem.* **31**, 308–314
(32) Turrell, E. S. & Robinson, S. (1942) The acid–base equilibrium of the blood in exercise. *Amer. J. Physiol.* **137**, 742–745
(33) Krebs, H. A. (1973) Pyridine nucleotides and rate control. *Symp. Soc. Exp. Biol.* **27**, 299–318
(34) Veech, R. L., Raijman, L. & Krebs, H. A. (1970) Equilibrium relations between the cytoplasmic adenine nucleotide system and nicotinamide–adenine nucleotide system in rat liver. *Biochem. J.* **117**, 499–503
(35) Stubbs, M., Veech, R. L. & Krebs, H. A. (1972) Control of the redox state of the nicotinamide–adenine dinucleotide couple in rat liver cytoplasm. *Biochem. J.* **126**, 59–65
(36) Burton, K. (1957) Free Energy Data of Biological Interest. *Ergeb. Physiol. Biol. Chem. Exp. Pharmakol.* **49**, 275–285
(37) Bassham, J. A. & Krause, G. H. (1969) Free energy changes and metabolic regulation in steady-state photosynthetic carbon reduction. *Biochim. Biophys. Acta* **189**, 207–221
(38) Mendel, B., Engel, W. & Goldscheider, I. (1925) Über den Milchsäuregehalt des Blutes unter physiologischen und pathologischen Bedingungen. *Klin. Wochenschr.* **4**, 262–263; 306–307; 542–544; 804–806
(39) Mendel, B. & Baruch, M. (1926) Über den Milchsäuregehalt des Blutes unter physiologischen und pathologischen Bedingungen. *Klin. Wochenschr.* **5**, 1272–1273
(40) Jervell, O. (1928) Investigation of the concentration of lactic acid in blood and urine under physiologic and pathologic conditions. *Acta Med. Scand. Suppl.* **24**, 1–135
(41) Eggleton, M. G. & Evans, C. L. (1930) The lactic acid content of the blood after muscular contraction under experimental conditions. *J. Physiol. (London)* **70**, 269–293
(42) Ross, B. D., Hems, R. & Krebs, H. A. (1967) The rate of gluconeogenesis from various precursors in the perfused rat liver. *Biochem. J.* **102**, 942–951
(43) Woods, H. F. & Krebs, H. A. (1971) Lactate production in the perfused rat liver. *Biochem. J.* **125**, 129–139
(44) Berry, M. N. (1967) The liver and lactic acidosis. *Proc. Roy. Soc. Med.* **60**, 1260–1262
(45) Huckabee, W. E. (1961) Abnormal resting blood lactate. II. Lactic acidosis. *Amer. J. Med.* **30**, 840–848

(46) Amatuzio, D. S. & Nesbitt, S. (1950) Study of pyruvic acid in blood and spinal fluid of patients with liver disease with and without hepatic coma. *J. Clin. Invest.* **29**, 1486–1490
(47) Seligson, D., Waldstein, S. S., Giges, B., Meroney, W. H. & Sborov, V. M. (1953) Some metabolic effects of ethanol in humans. *Clin. Res. Proc.* **1**, 86
(48) Mulhausen, R., Eichenholz, A. & Blumentals, A. (1967) Acid base disturbances in patients with cirrhosis of the liver. *Medicine (Baltimore)* **46**, 185–189
(49) Alberti, K. G. M. M. (1972) Gluconeogenesis in liver disease. *Conn. Med.* **36**, 568–571
(50) Record, C. O., Alberti, K. G. M. M., Williamson, D. H. & Wright, R. (1973) Glucose tolerance and blood metabolite changes in human viral hepatitis. *Clin. Sci.* **45**, 677–690
(51) Soffer, L. J., Dantes, D. A. & Dobotka, H. (1937) Sodium-d-lactate blood clearance as test of liver function. *Proc. Soc. Exp. Biol. Med.* **36**, 692–693
(52) Cohn, C. (1942) Sodium-d-lactate tolerance as a test of hepatic function. *Arch. Int. Med.* **70**, 829–835
(53) Bock, A. V., Dill, B. D. & Edwards, H. T. (1932) Lactic acid in the blood of resting man. *J. Clin. Invest.* **11**, 775–788
(54) Eldridge, F. & Salzer, F. (1966) Effect of respiratory alkalosis on blood lactate and pyruvate in humans. *J. Appl. Physiol.* **22**, 461–469
(55) Chamberlain, J. H. & Lis, M. T. (1968) Observations of blood lactate and pyruvate levels and excess lactate production during and after anaesthesia with and without hyperventilation. *Brit. J. Anaesth.* **40**, 315–322
(56) Sykes, H. K. & Cooke, P. H. (1965) The effects of hyperventilation on 'excess lactate' production during anaesthesia. *Brit. J. Anaesth.* **37**, 372–379
(57) Zborowska-Sluis, D. T. & Dossetor, J. B. (1967) Hyperlactatemia of hyperventilation. *J. Appl. Physiol.* **22**, 746–755
(58) Berry, M. N. & Scheuer, J. (1967) Splanchnic lactic acid metabolism in hyperventilation, metabolic alkalosis and shock. *Metab. Clin. Exp.* **16**, 537–556
(59) Cournand, A., Riley, R. L., Bradley, S. E., Breed, E. S., Nobel, R. O., Lauson, H. D., Gregersen, M. I. & Richards, D. W. (1943) Studies of circulation in clinical shock. *Surgery* **13**, 964–995
(60) Beatty, C. H. (1945) Effect of hemorrhage on lactate/pyruvate ratio and arterio-venous differences in glucose and lactate. *Amer. J. Phsyiol.* **143**, 579–588
(61) Seligman, A. M., Frank, H. A., Alexander, B. & Fine, J. (1947) Traumatic shock; carbohydrate metabolism in hemorrhage shock in dog. *J. Clin. Invest.* **26**, 535–546
(62) Ballenger, W. F., Vollenweider, H., Pierucci, L. & Templeton, J. Y. (1961) Anaerobic metabolism and metabolic acidosis during cardiopulmonary bypass. *Ann. Surg.* **153**, 499–506
(63) Ballenger, W. F., Vollenweider, H., Pierucci, L. & Templeton, J. Y. (1961) Acidosis of hypothermia. *Ann. Surg.* **154**, 517–523
(64) Broder, G. & Weil, M. H. (1964) Excess lactate: an index of reversibility of shock in human patients. *Science* **143**, 1457–1459
(65) Peretz, D. I., Schott, H. M., Duff, J., Dossetor, J. B., MacLean, L. D. & McGregor, M. (1964) The significance of lacticacidemia in the shock syndrome. *Ann. N.Y. Acad. Sci.* **119**, 1133–1141
(66) Alberti, K. G. M. M. & Hockaday, T. D. R. (1972) Blood lactic and pyruvic acids in diabetic coma. *Diabetes* **21**, Suppl. **1**, 350
(67) Watkins, P. J., Smith, J. S. & Fitzgerald, M. G. (1969) Lactic acidosis in diabetes. *Brit. Med. J.* **I**, 744–747
(68) Pitts, R. (1968) *Physiology of the Kidney and Body Fluids*, 2nd edn., p. 84, Year Book Medical Publishers Inc., Chicago
(69) Ram, B. K. & Ahuja, M. M. (1970) A study of intermediary metabolism in different clinical types of diabetes mellitus in India (with reference to pyruvate, lactate glycerol and plasma insulin). *Indian J. Med. Res.* **58**, 456–467
(70) Lyngsoe, J., Bitsch, V. & Trap-Jenson, J. (1972) Influence of phenformin on fat and lactate metabolism and insulin production in starved normal subjects. *Metab. Clin. Exp.* **21**, 179–186

(71) Varma, S. K., Heaney, S. J., Whyte, W. G. & Walker, R. S. (1972) Hyperlactataemia in phenformin-treated diabetes. *Brit. Med. J.* **I**, 205–206
(72) Toews, C. J., Kyne, J. L., Connon, J. J. & Cahill, G. F. (1970) The effect of phenformin on gluconeogenesis in isolated perfused rat liver. *Diabetes* **19**, 368
(73) Connon, J. J. (1973) A differential action of phenformin in normal and diabetic rat liver. *Diabetologia* **9**, 47–49
(74) Woods, H. F. & Alberti, K. G. M. M. (1972) Dangers of intravenous fructose. *Lancet* **ii**, 1354–1356
(75) Woods, H. F., Eggleston, L. V. & Krebs, H. A. (1970) The cause of hepatic accumulation of fructose 1-phosphate on fructose loading. *Biochem. J.* **119**, 501–510
(76) Mäenpää, P. H., Raivio, K. O. & Kekomäki, M. P. (1968) Liver adenine nucleotides: fructose-induced depletion and its effect on protein synthesis. *Science* **161**, 1253–1254
(77) Raivio, K. O., Kekomäki, M. P. & Mäenpää, P. H. (1969) Depletion of liver adenine nucleotides induced by D-fructose. Dose dependence and specificity of the fructose effect. *Biochem. Pharmacol.* **18**, 2615–2624
(78) Schumer, W. (1971) Adverse effects of xylitol in parenteral alimentation. *Metab. Clin. Exp.* **20**, 345–347
(79) Woods, H. F. & Krebs, H. A. (1973) Xylitol metabolism in the isolated perfused rat liver. *Biochem. J.* **134**, 437–443
(80) Oei, T. L. (1962) Hexose monophosphate, pyruvate and lactate in the peripheral blood in glycogen-storage disease type I. *Clin. Chim. Acta* **7**, 193–198
(81) Howell, R. R., Ashton, D. M. & Wyngaarden, J. B. (1962) Glucose 6-phosphatase deficiency glycogen storage disease. *Pediatrics* **29**, 553–565
(82) Mason, H. H. & Andersen, D. H. (1955) Glycogen disease of the liver (von Gierke's disease) with hepatoma. Case report with metabolic studies. *Pediatrics* **16**, 785–800
(83) Sokal, J. E., Lowe, C. H., Sarcione, E. J., Mosovich, L. L. & Doray, B. H. (1961) Studies of glycogen metabolism in liver glycogen disease (von Gierke's disease): six cases with similar metabolic abnormalities and response to glucagon. *J. Clin. Invest.* **40**, 364–374
(84) Brown, R. E., Madge, G. E. & Mamunes, P. (1972) Chronic lactic acidosis in infancy. *Arch. Pathol.* **94**, 192–196
(85) Hartmann, A. F., Sr., Wohltmann, H. J., Purkerson, M. L. & Wesley, M. E. (1962) Lactate metabolism: studies of a child with a serious congenital deviation. *J. Pediat.* **61**, 165–180
(86) Sussman, K. E., Alfrey, A., Kirsch, W. M., Zweig, P., Felig, P. & Messner, F. (1970) Chronic lactic acidosis in an adult. *Amer. J. Med.* **48**, 104–112
(87) Delvin, E., Scriver, C. R. & Neal, J. L. (1974) Pyruvate carboxylase in human liver. Apparent loss of a component of catalytic activity in a form of lactic acidosis with hypoglycaemia. *Biochem. Med.* **10**, 97–106
(88) Erickson, R. J. (1965) Familial infantile lactic acidosis. *J. Pediat.* **66**, 1004–1016
(89) Haworth, J. C., Ford, J. D. & Younoszai, M. K. (1967) Familial chronic acidosis due to an error in lactate and pyruvate metabolism. *Can. Med. Ass. J.* **97**, 773–779
(90) Israels, S., Haworth, J. C., Gourley, B. & Bord, J. D. (1964) Chronic acidosis due to an error in lactate and pyruvate metabolism. *Pediatrics* **34**, 346–356
(91) Nordio, S., Lamedica, C. M., de Pra, M. & Trovini, G. C. (1963) Iperlattacidemia idiopatica del lattante? *Minerva Pediat.* **15**, 1068–1073
(92) Schärer, K., Marty, A. & Mühlethaler, J. P. (1968) Chronic congenital lactic acidosis. A fatal case with hyperphosphatemia and hyperlipemia. *Helv. Pediat. Acta* **23**, 107–127
(93) Skrede, S., Strömme, J. H., Stokke, O., Lie, S. O. & Eldjarn, L. (1971) Fatal congenital lactic acidosis in two siblings. *Acta Pediat. Scand.* **60**, 138–145
(94) Worlsey, H. E., Brookfield, R. W., Edward, J. S., Noble, R. L. & Taylor, W. H. (1965) Lactic acidosis with necrotizing encephalopathy in two sibs. *Arch. Dis. Child.* **40**, 492–501
(95) Carlström, B., Myrback, K., Holmin, N. & Carsson, A. (1939) Biochemical studies on B_1-avitaminosis in animals and man. *Acta Med. Scand.* **102**, 175–213

(96) Elsom, K. O., Lukens, F. D. W., Montgomery, E. H. & Jonas, L. (1940) Metabolic disturbances in experimental human vitamin B deficiency. *J. Clin. Invest.* **19**, 153–161
(97) Stotz, E. & Bessey, O. A. (1943) The blood lactate pyruvate ratio relation and its use in experimental thiamine deficiency in pigeons. *J. Biol. Chem.* **143**, 625–631
(98) Kreisberg, R. A. (1974) Lactic acidosis and hyperlactataemia. *Lancet* **i**, 1351
(99) Cohen, R. D. & Woods, H. F. (1975) *Clinical and Metabolic Aspects of Latic Acidosis*, Blackwells Scientific Publications, Oxford in the press

Essays Med. Biochem. **1**, 105–131
Printed in Great Britain

Bence-Jones Proteins

By J. R. HOBBS

Department of Chemical Pathology, Westminster Medical School, London SW1P 2AR, U.K.

Historical

Thanks to Clamp (1) it is now clear that the urine of Mr. Thomas McBean, aged 45 years, was personally examined by Dr. William MacIntyre, a 53-year-old Harley Street physician, who observed a precipitate which appeared at 40–58°C when the weakly acidified urine was warmed and which disappeared at 100°C only to reappear on cooling. Dr. MacIntyre had been called in by the patient's own doctor, Dr. Thomas Watson, on October 30th, 1845. Next day a specimen of the patient's urine was sent to Dr. Henry Bence Jones, a 31-year-old physician at St George's Hospital, London, who was also an able pathologist. (He later became an F.R.S.) Bence Jones confirmed and extended MacIntyre's observations which led to the protein, which he called 'hydrated deutoxide of albumen', bearing his name (hyphenated when used as an adjective).

Today we continue to use the term Bence-Jones protein because this is how it is best known, but the mistaken conclusion that it is derived from albumin probably retarded progress in this field for 100 years. Interestingly, Bence Jones never found another such protein. This was probably not so much due to their rarity as to the very poor detection rate for random heat tests (see below).

Bence Jones estimated the concentration of the protein in urine at nearly 70 g/litre. The highest concentration we have observed in 1800 patients is 30 g/litre.

Mr. McBean died on January 1st, 1846, and it was Dr. John Dalrymple of the Moorfields Eye Hospital who first recognizedat the post mortem severe involvement of the axial skeleton with sparing of the long bones. The affliction was at that time called 'mollities et fragilitas ossium' and it was not until 1873 when von Rustizky introduced, in German, the term 'multiple myeloma' (2) that the bony origin of the tumours underlying the disease became clear.

In 1889 Kahler (3) reported the simultaneous presence of multiple myelomata and Bence-Jones proteinuria. The whole history of myelomatosis is well reviewed by Snapper & Kahn (4).

Bence-Jones protein is of especial interest to oncologists as it was the first tumour product to be recognized. Several other tumour products are now well-known and established as an important means of monitoring cancers (5, 6).

Definitions

Bence-Jones proteins are monoclonal light chains. The light chains of immunoglobulins are polyclonal and belong to two main classes; 64% are κ light chains and 36% are λ light chains. When monoclonal, light chains are referred to as Bence-Jones proteins. Although the methods were not available at the time to confirm it, there can be no doubt that the proteinuria originally found in Mr. McBean by Dr. MacIntyre were also due to monoclonal light chains.

Many years of work by Putnam and his colleagues (7) has established that light chains isolated from immunoglobulins can show the same solubility characteristics as Bence-Jones proteins and that Bence-Jones proteins are the products of synthesis *de novo* in the body rather than breakdown products of circulating immunoglobulins. The monoclonal concept which follows from Burnet's theory that one cell makes only one antibody is now established for over 98% of plasma cells (8). The few exceptions to this general rule probably represent either switch cells or overlapping specificity.

Switch cells are cells which synthesize the same V_H group (i.e. the variable portions of the immunoglobulin heavy chain which are responsible for much of the antibodies' specificity), but switch (see Fig. 1) from one constant region, $C\mu$ (e.g. that for IgM*), to another, $C\gamma$ (e.g. that for IgG), so that subsequent radioimmunoelectrophoresis reveals two distinct antibodies (9).

The concept of overlapping specificity derives from the theory of Talmage (10) which permits a single antibody to react with two indicator targets because of their sharing of an antigenic overlap. Within such provisions, we can accept that in general one cell makes only one antibody and usually incorporates only one class of light chain (11, 12). A few observations have, however,

* Abbreviations: IgM, immunoglobulin M; IgG, immunoglobulin G; IgD, immunoglobulin D; IgE, immunoglobulin E.

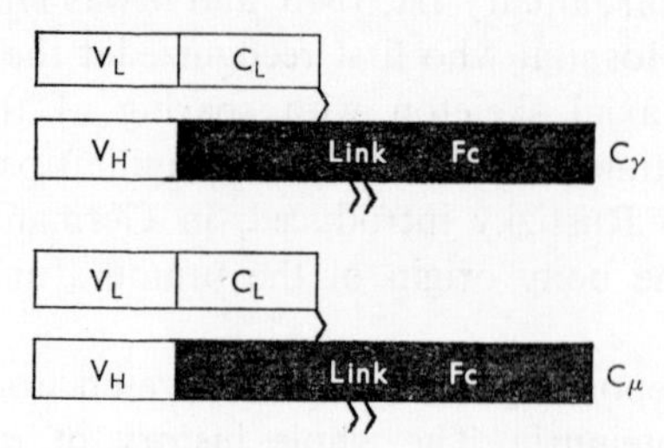

Fig. 1. *Illustration of how two related subclones can synthesize immunoglobulins differing only in the nature of their C_H fraction, presumably as a result of a switch from $C\mu$ to $C\gamma$ coding*

been made (13, 14) in which both κ and λ light chains were observed inside the same plasma cell (0.3–8.0% of them!).

Detection of Bence-Jones Proteinuria

Detection of Bence-Jones proteinuria is important in the diagnosis of myelomatosis as well as being of intrinsic interest to biochemists and immunologists. Unfortunately, many of the tests that have been used throughout the years are less than completely satisfactory.

The heat test was the first to be used and many variants of it have been introduced doubtless because of its poor reproducibility. In a complex form of the test introduced by Bernier & Putnam (15) the urine is buffered to pH 4.9, sealed in a glass capillary tube and heated in oil to 120°C. Even so, no less than 22 out of 66 urine samples known to contain Bence-Jones protein at a concentration of 0.15 g/litre or more failed to give a positive response by producing a protein precipitate at temperatures up to 59°C (16). In an even larger number (51 out of 66 or 78%) the precipitate that did form failed to redissolve at 120°C probably because other proteins were also present.

The simpler three-tube test for Bence-Jones proteinuria described by Harrison (17), in which the urine is filtered while still at 100°C and observed to see whether Bence-Jones protein precipitates out on cooling, gave an equally high percentage of false negative results. Moreover, urine containing no Bence-Jones protein but excess of α_1 globulins, which is common in patients with cancer, burns and severe inflammatory states, may also produce a precipitate before the temperature reaches 58°C. Thus, not only does the heat test miss one-third of the cases of Bence-Jones proteinuria, but it also produces false positive results and therefore cannot be considered wholly specific.

The Heller test performed by layering urine over concentrated nitric acid and observing whether a precipitate forms at or above the liquid junction was negative in 15 out of 79 (19%) urine samples known to contain Bence-Jones protein at a concentration greater than 0.05 g/litre. The sulphosalicylic acid test and trichloroacetic acid test were also falsely negative in 6 and 5% of the samples respectively. An even larger percentage (31%) of false negatives were obtained by the propan-1-ol test introduced by Jirgensons *et al.* (18). Albustix, a paper 'stick' method for detecting protein in urine, is of little value in screening for Bence-Jones proteinuria, and with the toluenesulphonic acid test of Cohen & Raducha (19) no less than 52% of 79 urine samples known to contain Bence-Jones protein were falsely negative.

The test for Bence-Jones proteinuria, described by Bradshaw (20), in which urine is carefully layered over concentrated hydrochloric acid is positive in 95% of cases (see Plate 1*a*) including those in which the concentration of Bence-Jones protein is as low as 0.01 g/litre. Because it chiefly demonstrates the presence of excess of globulin in the urine, the Bradshaw test is not specific for Bence-Jones protein. Nevertheless, it does represent a cheap and effective screening test.

The best method of seeking to demonstrate Bence-Jones proteinuria is electrophoresis of the concentrated urine. Without prior concentration, and using the routine cellulose acetate electrophoretic technique (21) in which 0.3 μl of fluid is applied across a 1 cm strip and stained with Ponceau-S or Naphthalene Black 10B (22), protein bands can be clearly detected at concentrations of about 2 g/litre or more. To demonstrate concentrations of Bence-Jones protein in the region of 0.01 g/litre, the urine must first be concentrated at least 200 times. Correspondingly, a lower concentration is required when Bence-Jones proteinuria is heavier.

We favour the collodion thimble illustrated in Fig. 2 to achieve satisfactory urinary concentration. The urine must first be pre-filtered through a suitable ultrafilter to remove particles which might clog the collodion thimble. This step also serves to remove bacteria which may be capable, after concentration, of destroying all of the protein in the urine within 4 h as did indeed happen in earlier days in our laboratory.

Ultrafiltration can conveniently be carried out on centrifuged or filtered urine by using a 20 ml syringe connected to a Swinnex 25 mm filter holder containing a 0.2 μm filter. Urine pushed through the filter directly into a collodion thimble will then concentrate very rapidly and 10 ml of urine can be reduced to 300 μl in about 1–2 h. Moreover, with pre-filtration the collodion

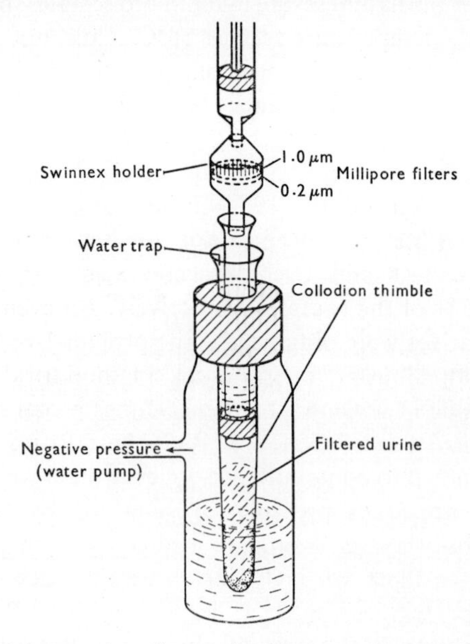

Fig. 2. *Illustration of the use of a collodion thimble to concentrate carefully prefiltered urine*

EXPLANATION OF PLATE I(*a*)

Bradshaw's test

The urine on the left, containing Bence-Jones protein, was layered over conc. HCl and has produced a white precipitate at the junction. The specimen on the right was from a patient with nephrotic syndrome and contained globulins but no Bence-Jones protein. Contrast the results with that of normal urine in the centre. Bradshaw's test can readily detect a slight excess of globulins.

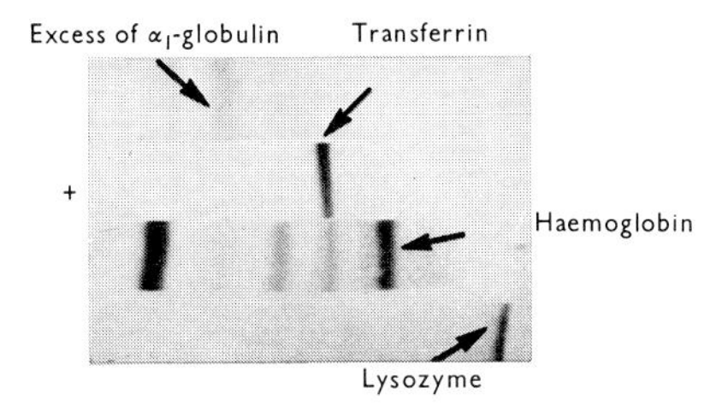

EXPLANATION OF PLATE I(*b*)

Electrophoretic strips obtained from four samples

The Plate shows that distinct narrow protein bands in the globulin region are usually due to Bence-Jones protein (not shown) or excess of α_1-globulin, haemoglobulin (or myoglobin) or, in a post-γ position, lysozyme. In urine, transferrin only occurs with albumin but here a pure sample is shown for the purpose of illustration.

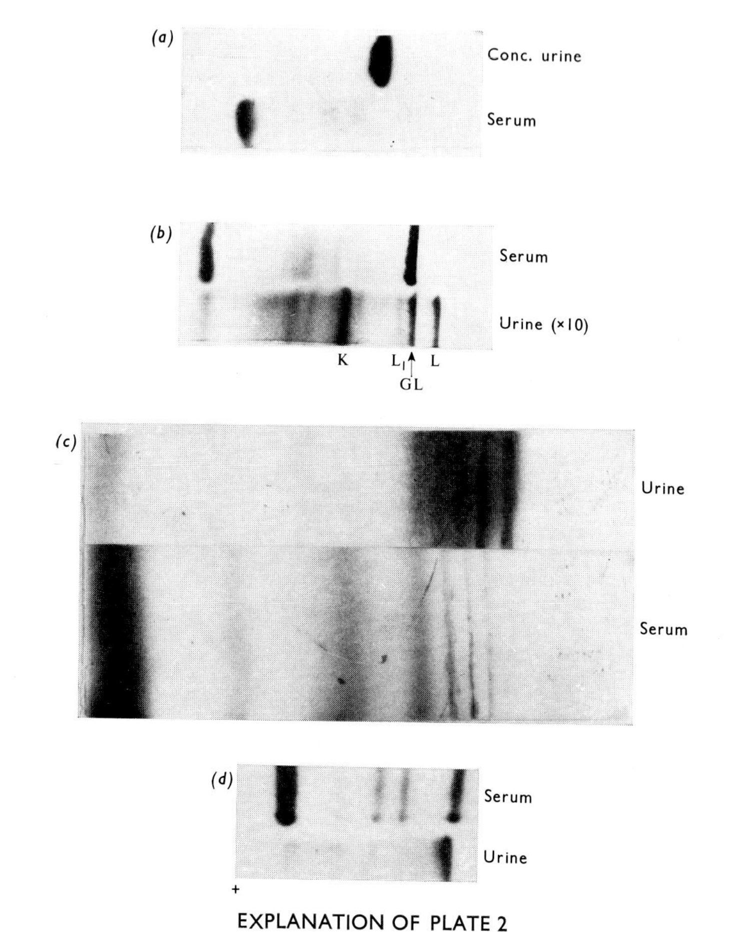

EXPLANATION OF PLATE 2

Electrophoretograms of serum and urine from patients with Bence-Jones proteinuria

(*a*) The single protein band in the urine electrophoretogram has no counterpart in the serum and is a monomer of Bence-Jones protein subsequently proved by immunoelectrophoresis. (*b*) One of the three sharp protein bands in the urine electrophoretogram is opposite the serum G_{λ} paraprotein and is present in amounts, relative to albumin, appropriate for the whole immunoglobulin. The bands on either side of it have no serum counterpart. They were of low molecular weight and were subsequently typed as λ (related to the $G\lambda$) and κ (from a second unrelated clone) Bence-Jones proteins. Biclonal Bence-Jones proteinuria such as this occurs in less than 2% of malignant immunocytomata. (*c*) This shows a harmonic series of urine bands of Bence-Jones proteins with increasing concentration in the opposite direction to those in the serum. The monomer on the right filters best, the dimer, trimer and tetramer (on the left) are increasingly retained in the serum. (*d*) In the urine electrophoretogram shown the γ band is associated with comparatively little albumin (i.e. it is of low molecular weight) but with an excessive proteinuria of α, β and γ mobilities typical of tubular proteinuria.

J. R. HOBBS

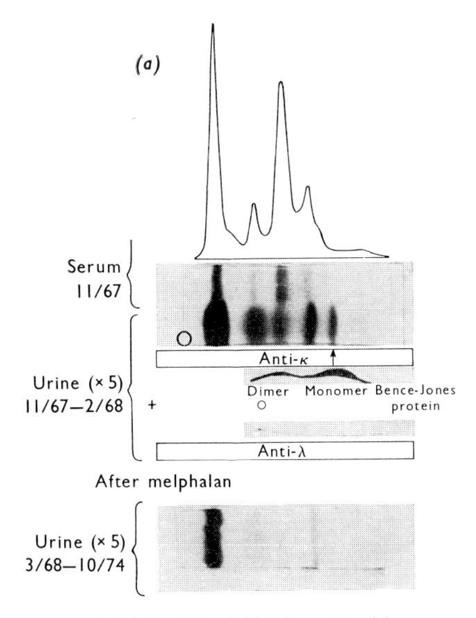

EXPLANATION OF PLATE 3(*a*)

Electrophoretograms of urine and serum from a patient with Bence-Jones proteinuria

The protein band marked with an arrow in the upper urine strip had no serum counterpart and was a κ monomer associated with amyloid nephropathy. It could only have been detected in this fashion. The bottom electrophoretogram shows that after treatment with melphalan, the Bence-Jones protein became undetectable, and renal function improved.

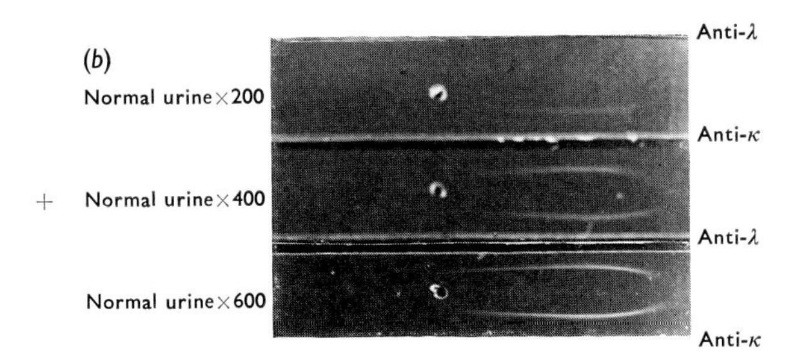

EXPLANATION OF PLATE 3(*b*)

Immunoelectrophoresis of normal urine against anti-κ and anti-λ antisera

As normal urine which has been concentrated × 600 is diluted (read from below upwards) its reaction with anti-λ is lost and the precipitin line against anti-κ antiserum shortens (top strip). This could easily be mistaken for a narrow reaction against κ chain only, but a narrow band would not be seen at the appropriate position in the simple protein electrophoretogram.

J. R. HOBBS

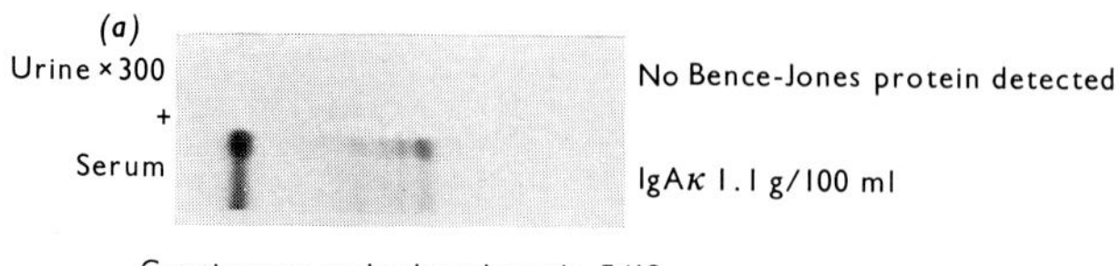

Continuous cyclophosphamide 5/12

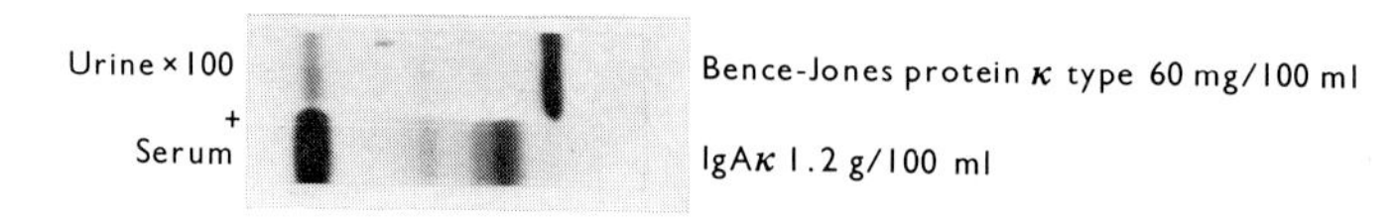

EXPLANATION OF PLATE 4(*a*)

'Bence-Jones escape'

Bence-Jones proteinuria appeared for the first time during a relapse occurring in the course of continuing cytotoxic therapy. The light chain (lower half of the Plate) was identical with that of the parent IgAκ paraprotein.

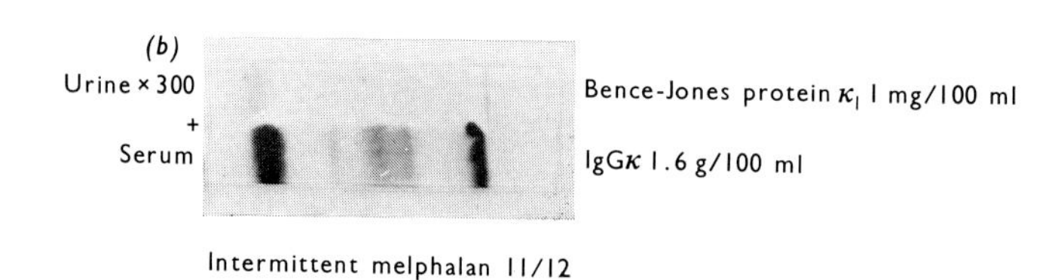

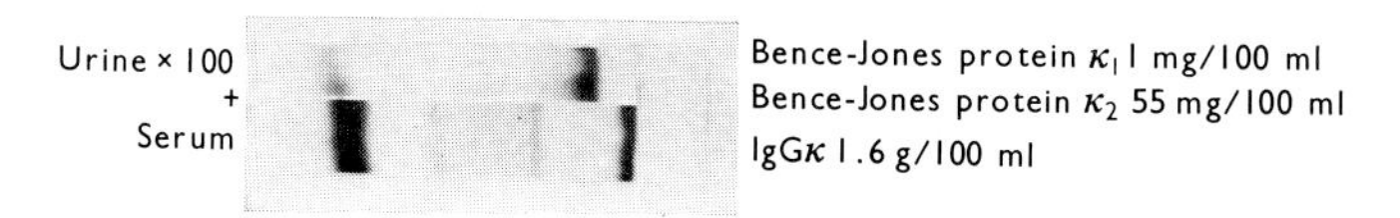

EXPLANATION OF PLATE 4(*b*)

'Mutation escape'

A second monomer of κ Bence-Jones protein appeared (lower panel) during relapse in a patient treated with melphalan. This Bence-Jones protein (κ_2) showed a selective and faster increase than the parent clone of IgGκ+κ_1.

J. R. HOBBS

thimbles do not clog and can be re-used up to ten times which is economical.

If electrophoresis buffer is used to surround the thimble during concentration the urinary concentrate will be dialysed ready for electrophoresis at the conclusion of this stage. As the larger molecules collect at the edges of the collodion thimble they form their own density gradient which is effectively self-wiping so that by the end of the concentration between 90 and 94% of ^{131}I-labelled albumin (molecular weight 70000) or 85 and 91% of ^{131}I-labelled Bence-Jones monomer (molecular weight 22000) can be recovered from the tip of the tube without washing its sides.

The urinary concentrate is best electrophoresed alongside the patient's own serum which enables the investigator to utilize clearance concepts in the interpretation of the strips (see below). The examples shown in Plates 1(*b*), 2 and 3(*a*) show how readily such combined strips can guide further investigation.

Plates 1(*b*), 2(*a*) and 2(*b*) show the simplest findings to interpret. Because there are no counterparts of the urinary protein band in the serum they can safely be assumed to have been completely cleared by the kidney, i.e. they are of low molecular weight. The only urinary proteins that could be confused with these Bence-Jones proteins by virtue of their mobility and concentration, are myoglobin and haemoglobin. Both of these proteins are highly coloured and consequently easily recognizable directly in the urinary concentrate and can be confirmed spectroscopically. In the case of a Bence-Jones protein with marked post-γ mobility, confusion could be caused by lysozyme.

Plates 2(*c*) and 2(*d*) show that even when the abnormal protein bands in the serum and urine have identical electrophoretic mobility, the density of the urine band relative to that of albumin may indicate that it is of lower molecular weight (i.e. <70000), and probably therefore a Bence-Jones protein.

Plate 2(*c*) shows, in addition, that the presence of Bence-Jones polymers can even be surmised simply by electrophoretic analysis of the urine.

Plates 2(*d*) and 3(*a*) show that associated renal damage can obscure definition. Nevertheless, with care, Bence-Jones proteinuria can still be discerned. It is in situations such as these that electrophoresis alone is capable of detecting Bence-Jones proteinuria. Heat tests are hopeless, the Bradshaw test is bound to be positive because of the excess of globulins, and immunoelectrophoresis will inevitably detect excess of κ and λ in the urine.

It cannot be overstressed that in order to diagnose Bence-Jones proteinuria a narrow band which indicates its monoclonal nature must be seen on electrophoresis regardless of whatever other tests may show. The dye-binding of such bands as a percentage of the total of the strips can be use to derive the amount of Bence-Jones proteinuria from a total protein estimation (69).

Immunoelectrophoresis

Confirmation that a narrow protein band detected during electrophoresis of a urinary concentrate is indeed due to a Bence-Jones protein can be obtained by immunoelectrophoresis which can be used to demonstrate the presence of only one class of light chains (see Plates 1*b*, 2*a*, 2*b* and 2*c*). Reliable antisera are essential as one of the commonest reasons, in practice, for obtaining an arc against κ alone is the use of a poor anti-λ antiserum. Another is simply that there are usually twice as many κ chains as λ chains even when there is excessive excretion of immunoglobulins in patients with tubular proteinuria. Consequently, a urine concentrate may therefore contain enough κ chain but not enough λ chain to show up on immunoelectrophoresis. Moreover, as the concentration of antigen decreases the arc shortens and may eventually give the false impression of a 'monoclonal bow' (see Plate 3*b*). These potential sources of error emphasize the need to detect a narrow protein band in the regular protein electrophoretogram corresponding to the arc precipitated in the immunoelectrophoresis procedure. This illustrates, moreover, that immunoelectrophoresis alone is inadequate to establish a diagnosis of Bence-Jones proteinuria. The result shown in Plate 3(*b*) for example, could be obtained with almost any normal urine concentrate!

Some workers still advocate setting up multiple dilutions of the antisera and antigens in order to try and obtain, by approaching the ideal balance, a permanent arc of precipitate. This is costly and in practice we have found it more valuable to use a fixed antiserum concentration, usually neat, and to observe the development of arcs at ½, 1, 1½ and 3 h. In this way it is possible to record any arc that appears and subsequently disappears owing to an excess of either of the reactants. Because low-molecular-weight components such as Bence-Jones protein may produce precipitate across the 3 mm of agarose in 1½ h, whereas whole immunoglobulins rarely produce arcs within this time, it is possible that frequent inspection may provide information that cannot be obtained by less regular examination.

Other techniques

Sodium dodecyl sulphate–polyacrylamide has been skilfully used by Virella *et al.* (23) to aid in the diagnosis of Bence-Jones proteinuria but has disadvantages. Antiserum identification of the protein band cannot be carried out directly. Instead either duplicate tubes must be set up or the original gel cut in half before being placed in agarose for immunological identification. The technique of immunodiffusion (24, 25) enables simultaneous sizing and immunological identification of proteins to be carried out. It is possible by using Sephadex G-75 to establish the presence of light-chain monomers, dimers, tetramers or half-light chains. Half-molecules of immunoglobulin can also be identified in this way (26).

Methods of Isolation

Bence-Jones proteins can be isolated by various means from the urine of patients suffering from myelomatosis. Isolation is simplified by selecting a urine which contains a large amount of Bence-Jones protein and few additional proteins, demonstrable by electrophoresis. Many biochemists, following tradition, have used salt fractionation as their first step, mixing urine in equal parts with saturated ammonium sulphate. This has two great disadvantages. The first is that precipitation of the Bence-Jones protein is generally incomplete and unrepresentative; half-light chains, for example, remain in the supernatant. Secondly, the precipitate is often heavily contaminated. Instead of initial salt fractionation it is much better to first concentrate the urine and then dialyse it. This must be done under aseptic conditions, however, and cannot be achieved merely by adding azide to the urine. Prior ultrafiltration through millipore filters to remove bacteria is necessary. Subsequent work should be carried out at 4°C. Antiseptics, such as merthiolate, which complex to proteins, should be avoided.

It is generally possible, with visking tubing (e.g. 28/32) used under controlled vacuum (10lb/sq. in) to retain most of the proteins of molecular weight greater than 10000 (see Fig. 3). Most of the residual pigments and low-molecular-weight compounds can be removed by dialysis.

In this way solutions with a total protein content of 20–100g/litre can be prepared. Tubular proteins such as β_2-microglobulin, will always be

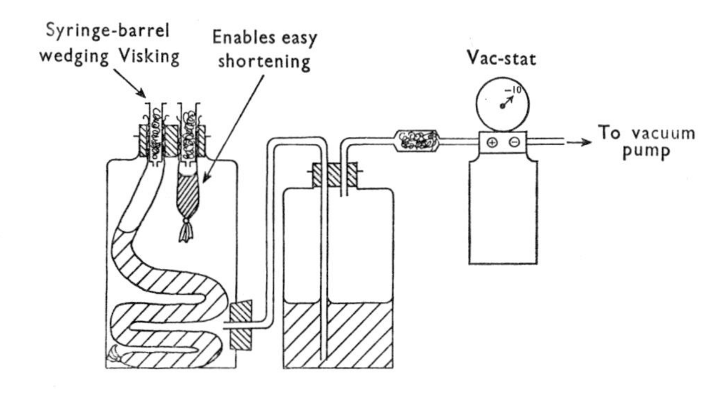

Fig. 3. *Illustration of a method for concentrating large quantities of urine*

The pre-filtered bacteria-free urine is protected by sterile cotton-wool plugs and the Visking tubing can easily be shortened (on the right) to facilitate harvesting the final concentrate. The concentration is best done at 4°C.

present and further separations should include at least one which is dependent on molecular size rather than on electrical charge alone. Mild reduction can also be used to dissociate complexes between light chains, especially κ, and other proteins (27). An alternative technique is to fractionate the urinary protein concentrate on a column of Sephadex G-50. This ensures that any dimer or monomer peaks can be collected. These can, in turn, be submitted to isoelectric focusing to obtain a final protein separation based on monoclonal electrophoretic mobility. A third method of obtaining pure Bence-Jones protein is to subject the urinary protein concentrate to sodium dodecyl sulphate–polyacrylamide-gel electrophoresis (28, 29, 23). Up to 25 mg of Bence-Jones protein can be isolated in this way.

If the isolated material is intended to be used to raise antisera it is helpful to add insolubilized antibodies against a concentrate of tubular proteinuria. The antibodies should be obtained from animals of the same species as that which will be used for immunization and although they will undoubtedly contain anti-(light chains) which will decrease the yield of Bence-Jones protein, they will also remove most of the contaminating proteins.

Antisera

Too many workers have been complacent in the past and have used commercial reagents which have been inadequately specified and have often been raised against a single Bence-Jones protein.

As will be explained later there are at least eight subclasses of light chains and, even in the constant regions, five variations occur. For this reason only antisera reacting with all the κ subclasses or with all of the λ subclasses should be used for routine detection. To ensure this, antigens used to produce antisera should be obtained from (i) a pool of known κ Bence-Jones proteins, (ii) a pool of known λ Bence-Jones proteins or (iii) normal light chains isolated from pooled IgG globulin and which can generally be prepared by DEAE-cellulose chromatography reasonably free from low-molecular-weight (tubular) proteins which usually contaminate pools (i) and (ii).

The harvested anti-(Bence-Jones protein) antisera should be adsorbed with immunoadsorbent-insolubilized proteins of pool (i) or (ii). This precaution avoids leaving immune complexes in the antisera, especially those against half-light chains and similar low-molecular-weight fragments (e.g. β_2-microglobulin) which are commonly present in pools (i) and (ii). The full procedure has been well described by Takahashi *et al.* (12).

Double-staining of single cells by an immunofluorescent technique, with antisera prepared according to these rigorous conditions, has not yet been reported. It may well be, therefore, that claims that κ and λ chains occur in the same plasma cell have been due to the use of inadequately adsorbed antisera. Indeed there is a 54% homology in the amino acid sequences of κ and λ chains.

Claims have been made that antisera can be produced which react only with 'free' light chains and not those 'bound' within immunoglobulin molecules (30). Although this is possible, problems due to idiotypes, molecular-weight variations (e.g. half-light chain to tetramer), half-molecules and especially tubular-protein contamination, make it highly unlikely that measurement of free light-chain content of biological fluids such as plasma and urine is specific.

Little reliance can be placed, at the present time, on immunochemical estimates of the plasma or urinary content of Bence-Jones protein or κ/λ ratios as they do not stand up to quality control examinations. In one survey, for example, the same urine sample was reported variously by different laboratories as containing from 0.3 to 31.5 g/litre of a Bence-Jones protein.

For clinical-chemistry purposes the best antisera are those that detect any κ or any λ Bence-Jones protein, whether 'free' or 'bound' as they enable the monoclonal structure of a suspected paraprotein to be verified.

Structure

Detailed reviews of the structure of Bence-Jones proteins are available elsewhere (31). Here it is intended only to summarize the information relevant to clinical chemistry.

There are, on the basis of amino acid-sequential analysis of their variable regions, at least three subclasses for κ light chains and five for λ light chains. The constant portion of the light chain on the other hand is deeply buried in intact immunoglobulin molecules and consequently less available for reaction with antibodies raised against them. It is important therefore for analytical work to use antisera that react with the variable determinants, presumably because these are more exposed since they are largely responsible for antigen-binding. Indeed some haptens, e.g. dinitrophenol, evoke largely κ antibodies, whereas others, e.g. pipsyl, evoke mainly λ antibodies (32). This selectivity probably explains the frequent failure of antisera raised against a single monoclonal Bence-Jones protein to detect light chains in intact monoclonal immunoglobulins of a different variable regional subclass, but nevertheless reacting, at least partially, with normal immunoglobulins since these contain representatives of the light-chain variable subclass against which the antiserum was raised.

Using antisera raised against a large pool of monoclonal light chains containing all of the κ and λ subclasses, I have experienced no difficulty in typing more than 1500 monoclonal immunoglobulins. Nevertheless, selective light-chain subclass antisera, and even antisera produced only against the variable region (33) may be invaluable in establishing the presence of 'switch-cell' clones (9) or closely related clones (34).

Selective light-chain subclass antisera can also be used to establish the relationship between Bence-Jones proteinuria and amyloidosis in an individual patient. The establishment of this relationship is one of the important contributions to knowledge to come from amino acid-sequencing studies of immunoglobulin. It has been demonstrated by Glenner *et al.* (35), for example, that some fractions, representing roughly 70% of purified amyloid fibrils, have amino acid sequences very similar to those of known Bence-Jones proteins.

Genetics

The hypothesis (36) that two genes are coded for immunoglobulin light chains; one gene for the variable (V, specificity) region, the other for the constant (C, common) region is now widely accepted. It seems likely that all individuals are coded for all of the light-chain variants, i.e. $V_{\kappa 1-3}$ and $V_{\lambda 1-5}$, although in a patient observed by R. Ballieux (unpublished work) there appeared to be a genetic absence of λ.

Synthesis

Immunoglobulins have been shown (37) to result from the balanced synthesis of heavy chains which take $2\frac{1}{2}$ min, and of light chains which take 1 min. Synthesis takes place on ribosomes of two different sizes (see Fig. 4) and is most rapid during the S phase and slowest during mitosis and the early G1 phase of the cell cycle (38).

After assembly, the intact immunoglobulin molecules move through the cisternae of the endoplasmic reticulum where carbohydrate is added to them. Eventually they reach the region of the Golgi apparatus where some final carbohydrate is added. Secretion occurs 30–40 min later. The delay may result from the fact that synthesis of the solubilizing light chains is confined to the Golgi membranes, whereas heavy-chain synthesis occurs on polyribosomes which are freely distributed throughout the cell (39).

Before 1958, Bence-Jones proteins were considered to be breakdown products of myeloma globulins. In that year Putnam & Miyake (40) gave [^{14}C]glutamate to a patient with multiple myelomatosis and showed that the Bence-Jones protein in his urine acquired a higher specific radioactivity than did the myeloma immunoglobulin in his plasma (Fig. 4). This finding implied synthesis *de novo* of the Bence-Jones protein. When subsequent work by other investigators showed that balanced synthesis of light and heavy chains occurs in well-differentiated plasma cells, I was led to postulate that

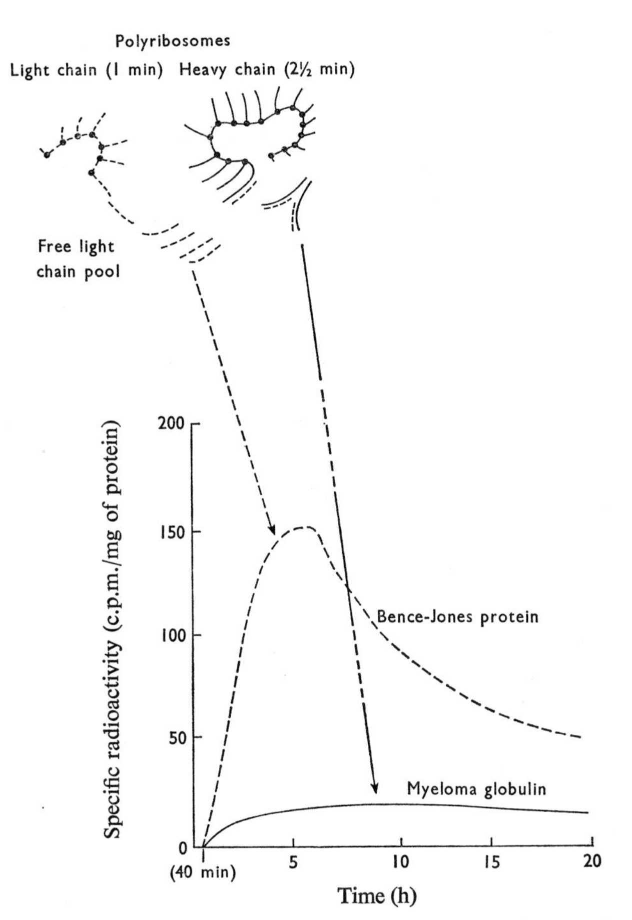

Fig. 4. *Diagrammatic demonstration that Bence-Jones protein cannot be a breakdown product of extracellular myeloma protein, but could be the result of excessive synthesis of free light chains*
The results are taken from Askonas & Williamson (37) and Putnam & Miyake (40). For further details see the text.

Bence-Jones-protein production in patients with myelomatosis can best be explained as the result of imbalanced synthesis by the undifferentiated plasma cells which characterize this disease. In 20% of patients with myelomatosis further dedifferentiation of plasma cells occurs and only light chains are synthesized and released (16, 41) or are deposited within the Golgi membranes as intracellular crystals (42). Such dedifferentiated myelomata tend to grow faster, occur in a younger age group (see Fig. 5), be more invasive insofar as more extensive bone and soft-tissue lesions are found, show primitive plasma cells in biopsies more often and carry a worse prognosis than other myelomata (43).

Half-light chains are found in the urine of 35% of the patients with these dedifferentiated myelomata (44). Since these largely represent the variable half (V_L) of the molecule, which is also presumed to be the first half to be synthesized, their presence provides further evidence of protein malsynthesis in the abnormal plasma (myeloma) cells.

The full clinical picture of myelomatosis is occasionally encountered (45) in which Bence-Jones-protein excretion in the urine is exceedingly low, sometimes amounting to as little as 2% of the average daily urinary excretion of 5.6g for patients with myelomatosis. Fluorescent-antibody studies of the myeloma cells in such patients revealed that their light-chain content is well below the usual, i.e. there is failure to produce even the usual quota of light chains. It can be concluded, therefore, that throughout the whole spectrum of myelomatosis an increase in biochemical dedifferentiation is paralleled by an increase in malignant dedifferentiation.

In some patients with myelomatosis and soft-tissue plasmacytoma, monoclonal products may be found only with great difficulty. In such cases biopsies can be studied by immunofluorescent or direct electrophoretic or immunoelectrophoretic techniques in order to establish the diagnosis. Immunoelectrophoresis can be carried out by inserting thin slices or a cell suspension of the biopsy material directly into slits or wells in an agarose slab. In this way, electrophoresis can lead to separation of proteins straight out of the cells (46, 47), and the demonstration of intracellular synthesis, possibly without secretion (48), even when other procedures have failed to reveal it. Nevertheless, even with these refined techniques some 1–2% of myelomatous tumours which are clearly derived from the plasma cell line and possess endoplasmic reticulum fail to synthesize recognizable heavy or light chains, i.e. they have become completely undifferentiated (45).

Studies of fresh tumours for synthesized surface markers (immunoglobulins of only one subclass, which are regenerated after their prior removal by trypsin treatment) reveal these on many lymphocytoid tumours, e.g. some 98% of chronic lymphatic leukaemia, some 60% of lymphosarcomata. Bence-Jones proteinuria can be detected with such tumours. Where such cells differentiate towards plasma cells the surface markers are lost, but they can also be lost during the evolution of the disease, presumably by further dedifferentiation (49).

Pathophysiology

Bence-Jones proteins are usually synthesized as monomers of molecular weight 22000, but readily undergo post-synthetic alterations. The commonest is dimer formation, usually through disulphide linkages.

Light-chain monomer is rapidly cleared through the renal glomerulus, at about 40% of the glomerular filtration rate. Light-chain dimer is also readily cleared but at only 20% of the glomerular filtration rate.

λ light chains polymerize more readily than κ chains. Consequently κ type Bence-Jones proteinuria generally contains mainly monomer but with some dimer, whereas λ-type Bence-Jones proteinuria often contains only dimer. As a result of polymerization to dimer, but partly also owing to an associated decrease in renal clearance, it is not unusual to find Bence-Jones proteins in the plasma (Bence-Jones proteinaemia) (50). Less commonly a closed light-chain tetramer (molecular weight 88000) usually of λ-type, can result in Bence-Jones protein being found only in the blood and not in the urine (51). There has even been one case of viscosity syndrome due to highly polymerized Bence-Jones proteinaemia.

Other kinds of post-synthetic alteration of light chains occur. These include complexing, especially of κ chains with prealbumin, albumin, α_1-antitrypsin (27) or transferrin. Detection of half-light chains where the amounts of V_L and C_L are equal suggests that post-synthetic proteolytic splitting has occurred. Malsynthesis is the probable explanation when the half-light chains consist only of V_L.

Turnover studies on Bence-Jones protein have been done mainly by using tracer injections in animals or patients with renal failure. From the former it is clear that $t_{\frac{1}{2}}$ of light chains is only 1.4 h. This can be prolonged to 9–17 h by nephrectomy, but is unaffected by ureteric ligation. It is concluded, therefore, that almost all of the Bence-Jones proteins administered are filtered at the glomerulus and metabolized by the kidney (52). The total capacity for such metabolism has not been assessed, however. It is known that in man renal tubular poisoning only induces proteinuria to the extent of about 1 g/day which may, therefore, represent the total capacity for reabsorption. In patients with no evidence of renal failure, Bence-Jones proteinuria (3–72 g/day) perfectly mirrors the serum paraprotein levels whether they fall or rise. I have inferred from this that the kidney's role in modification of their output of Bence-Jones protein is a minor one or that the reabsorptive capacity is readily saturated. Just why some Bence-Jones proteins damage the kidneys as evidenced by their association with azotaemia, albuminuria or tubular proteinuria (see Plate 2*d*) whereas others do not, despite large throughputs, is unknown. The readiness of some Bence-Jones proteins to co-precipitate in the tubules could possibly account for the renal failure (53) which is associated with myelomatosis in 10–15% of cases. This so-called myeloma-kidney is not specific to myelomatosis, however.

Although Morgan & Hammack (54) have dismissed intravenous pyelography as non-hazardous in patients with myelomatosis, Gross *et al.* (55) agree with me that such procedures should be avoided in patients with Bence-Jones proteinuria. Moreover, like E. F. Osserman (personal communication), I encourage a urinary fluid throughput of 3 litres/day, especially as hypercalcaemia, which is itself capable of producing kidney damage, is common in patients with myelomatosis.

Patients with Bence-Jones proteinuria often develop amyloid. Rarely they can develop (in less than 1% of cases) an acquired or 'adult' Fanconi syndrome (56) leading to acidosis and osteomalacia with subnormal serum levels of inorganic phosphate and calcium, a most unusual combination in cases of Bence-Jones proteinuria.

It is believed some particular types of Bence-Jones proteins are capable of inhibiting the normal metabolism of renal tubular cells.

Four examples of cryo-Bence-Jones proteins have been reported (57) and I have personally seen a fifth. In all of these cases light chains were found to aggregate like monoclonal cryoglobulins at low temperatures. They did not do so at body temperature, however, so that no symptoms referable to cryoprecipitation were observed, unlike in cryoglobulinaemia.

Are there normal counterparts to Bence-Jones proteins?

When normal urine is sufficiently concentrated, i.e. 400 times or more, both κ and λ chains can be identified (58) in the usual proportion of $2\kappa/1\lambda$ at levels of 2.5 mg/litre and show as diffuse γ-globulins on electrophoresis. They are polyclonal, therefore, and clearly distinguished from monoclonal light chains, i.e. Bence-Jones proteins. It is possible after even greater concentration to isolate half-light chains, which are probably the products of post-synthetic proteolysis since the amounts of V_L and C_L are equal.

In any situation in which there is increased glomerular-protein leakage or tubular damage, light chains are excreted in greater amounts than normal as the renal tubular reabsorptive capacity for proteins is relatively non-selective and readily saturated. More polyclonal light chains are excreted in the urine in conditions in which immunoglobulin turnover is increased. Thus in patients with chronic pyelonephritis or systemic lupus erythematosus, for example, light chains may show up on immunoelectrophoresis of urine which has only been concentrated $\times 50$ or less. The precipitates are long and smoothly graded in thickness, however, with κ usually showing up twice as easily as λ, quite unlike the monoclonal bows characteristics of Bence-Jones proteins (but see Plate 3*b*).

Proving the presence of Bence-Jones protein

Bradshaw's test offers a routine screening test for Bence-Jones protein in urine, but whenever the level of suspicion is high—in any suspected B-lymphocyte tumour, for example—fresh urine should be concentrated up to $\times 300$ and electrophoresed. It is not worth wasting time on immunoelectrophoresis unless a predominant narrow protein band is seen in the urine relative to the donor serum. If a narrow band is seen, however, it should be identified as being exclusively κ or λ with a monoclonal bow in the appropriate position (see Plate 3*a*) and unreactive with anti-(heavy chain)antisera. Occasionally, when there is anuria or a light-chain tetramer is suspected, it may be necessary to test the serum as well as the urine.

A new screening test for Bence-Jones proteinuria, namely rocket immunoselection, seems promising (59). In this technique neat urine is electrophoresed into antisera raised against the heavy chains γ, α and μ. Light chains present in the urine which are attached to heavy chains as immunoglobulins are precipitated while those which are not so attached continue onwards through a layer of ordinary agarose to reach a third, uppermost, layer containing antiserum against either κ or λ chains. This screen detects as little as 0.01 g of Bence-Jones protein per litre. Any κ or λ chains detected can be investigated further as already described.

In my own laboratory I always check the serum for the presence of IgD whenever I find 'only' Bence-Jones proteinuria, especially of the λ type. Any other unexplained serum band can be checked for IgE or other immunoglobulins.

Clinical Significance of Bence-Jones Proteinuria

Malignant immunocytomata

Table 1 summarizes the results of more than 10 years work in my own laboratory. It shows that Bence-Jones proteins were found in 660, i.e. 61 %, of 1086 patients who had paraproteins.

I have found neither paraproteins nor Bence-Jones proteins in patients with true Hodgkin's disease, although three patients had initially been given this diagnosis. In these three patients no Reed–Sternberg cells were found and they were subsequently rediagnosed as suffering from lymphosarcoma.

The reticulosarcomata were undifferentiated tumours and did not include five myelomata which terminated in similar fashion. All five patients with Bence-Jones proteinuria and cold haemagglutinin disease eventually developed frank lymphosarcoma after 3–15 years of typical haemolysis. Here the findings of Bence-Jones proteinuria seemed to forewarn of such an outcome (60).

Patients with IgM paraproteins can be classified according to whether symptoms due to plasma viscosity or symptoms due to tumour dominated the clinical picture. It can be seen that patients with frankly invasive malignant lymphomas have a higher incidence of Bence-Jones proteinuria than those who do not. Indeed, we have never demonstrated Bence-Jones proteinuria in patients with conditions which have not shown malignant invasion over a 5-year observation period. This includes patients with lichen myxoedematosus,

Table 1. *Incidence of Bence-Jones protein in* 1410 *patients in whom thorough examinations have been made*

Number of patients	Number with para-proteins	Final diagnosis	Number with Bence-Jones proteinuria	Percentage of patients with that diagnosis showing Bence-Jones proteinuria
708	699	Myelomatosis	488	69
34	31	Soft-tissue plasmacytoma	21	60
104	18	Chronic lymphatic leukaemia	16	15
46	46	Waldenstrom's IgM viscosity	36	78
53	53	Malignant IgM lymphoma	49	92
84	0	Reed-Sternberg Hodgkin's disease	0	0
41	3	Reticulosarcoma	3	7
49	49	Primary cold haemagglutinin disease	5	10
31	1	Giant-follicular lymphoma	1	3
94	20	Other lymphoma	18	19
8	8	Lichen myxodematosus	0	0
7	7	Transient paraproteins	0	0
76	76	Benign paraproteins (>5 years)	0	0
75	75	Unknown (<5 years)	23	—
1410	1086*		660	

* Includes 155 with only Bence-Jones proteins.

Table 2. *Incidence of Bence-Jones proteinuria in myelomatosis*

The results exclude 22 biclonal cases.

Paraprotein class	Number of patients	Number with Bence-Jones proteinuria	% with Bence-Jones proteinuria
IgG	406	244	60
IgA	171	122	71
Bence-Jones protein only	101	101	100
IgD	15	15	100
IgM	6	6	100
No paraprotein	9	0	0
Totals	708	488	69

transient and benign paraproteins and 44 of those with cold agglutinins. In 23 patients in whom Bence-Jones proteins have been demonstrated and the eventual final diagnosis is still awaited, observations have not yet been continued for 5 years.

Table 2 shows further details of patients in whom a firm diagnosis of myelomatosis was established by using Medical Research Council criteria, i.e. the presence of at least two of the following three criteria:

(i) discrete radiological lesions in the bones;
(ii) abnormal plasma cells in biopsy;
(iii) paraproteins in the plasma and/or urine.

It can be seen that all of the patients with IgD or IgM myelomata had Bence-Jones proteinuria.

Table 3. *Incidence of Bence-Jones proteinuria in soft-tissue plasmacytoma*

Paraprotein class	Number of patients	Number with Bence-Jones protein
IgG	6	3
½mol. Gκ	3	0
IgA	2	0
IgM	3	1
IgD	1	1
Bence-Jones protein only	16	16 (47%)
No paraprotein	3	0
Totals	34	21 (60%)

Table 4. *Expected frequency of Bence-Jones proteinuria in new patients annually per 50 million population*

Disease	Expected number of new patients	Expected percentage with Bence-Jones proteinuria	Number of patients with Bence-Jones proteinuria	Percentage of all new patients with Bence-Jones proteinuria
Myelomatosis	2400	70	1680	71
Soft-tissue plasmacytoma	100	60	60	3
Lymphosarcoma	1500	20	300	13
Reticulosarcoma	500	5	25	1
Waldenstrom's macroglobulinaemia	80	80	64	3
Chronic lymphatic leukaemia	1500	15	225	9
Totals	6080	34	2354	

Table 3 records the findings, before surgery, in 34 patients with soft-tissue plasmacytomata. It shows that in 47% of them Bence-Jones protein was the only paraprotein detected. Conversely, there was a tendency for patients with Bence-Jones proteins only and those with IgD myelomatosis to have involvement of the soft tissues (45).

Evidence for excretion of Bence-Jones proteins by patients with tumours of the B-lymphocyte line, apart from myelomatosis, derives from as long ago as 1909. In that year Decastello (61) recorded the presence of Bence-Jones proteinuria in a patient with typical chronic lymphatic leukaemia. More recent observation are those of Lindstrom *et al.* (62).

In Table 4 I have tried to estimate the number of diagnoses and biochemical findings that might be made if all of the patients in the U.K. suspected of harbouring tumours of the haemopoietic or lymphoid tissues had their urine properly examined.

Heremans introduced the term, immunocytoma, to describe tumours presumed to be of B-lymphocyte origin and to be capable of synthesizing immunoglobulins or fragments thereof. Immunocytomata can be further subdivided into benign and malignant (45) and it can be seen from Table 1 that the presence of Bence-Jones proteins in the urine is virtually synonymous with malignant immunocytoma. The diagnosis of malignant immunocytoma was established by independent criteria in no less than 637 of the 660 patients in whom Bence-Jones proteinuria was found. The remaining 23 patients have not been observed for long enough yet to exclude a similar diagnosis.

Very occasionally Bence-Jones protein is associated with a benign lesion and in a patient observed by E. F. Osserman (personal communication) the Bence-Jones proteinuria resolved spontaneously. Dammaco & Waldenstrom (63) claimed that Bence-Jones proteinuria may occur in patients with benign paraproteinaemia. It should be emphasized, however, that no less than six of their ten patients had been observed for less than 5 years. In the remaining four patients, Bence-Jones protein concentrations in the urine were said to be in the region of 17–58.4 mg/litre, but the method of measurement used by these authors is inaccurate at concentrations below 500 mg/litre, and the types of antisera used by them to prove the presence of free Bence-Jones proteins do not meet the criteria for specificity given above.

Notwithstanding, benign Bence-Jones proteinuria greater than 10 mg/litre probably does occur but is so rare that for practical purposes all such cases should be considered as indicative of a malignant condition until at least 5 years of observation have passed. Indeed, in one of our own patients 17 years elapsed

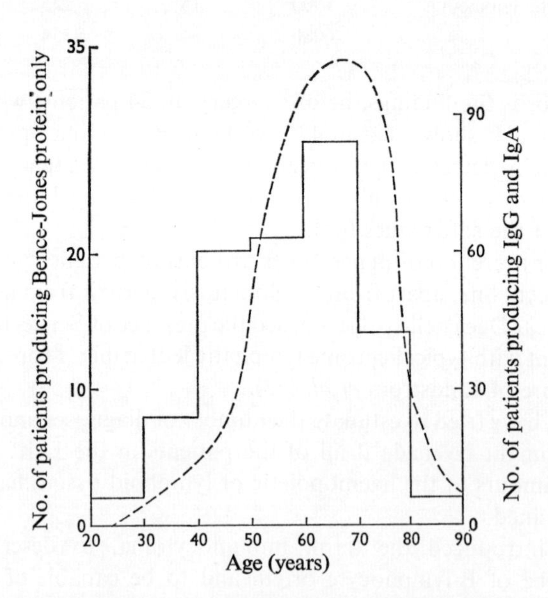

Fig. 5. *Frequency distribution by age of patients with myelomata producing whole immunoglobulins (IgG, IgA) (-----) or only Bence-Jones proteins (——)*

The latter account for most patients with paraproteins under 45 years of age.

Table 5. *κ/λ ratios for paraproteins according to class*

Number typed	Paraprotein class	κ/λ
237	IgG	2.04
84	IgA	1.63
24	Bence-Jones protein only	0.85
345	Subtotal	1.78
102	IgM	1.00
38	IgD	0.03

from the first detection of Bence-Jones proteinuria to development of clinical myelomatosis and eventual death from it. Throughout this period his daily urinary output slowly increased. In all other cases of initially unexplained or seemingly benign Bence-Jones proteinuria in which a diagnosis of malignancy was eventually made, it was made within 8 years of the first observation. As shown in Table 1, some patients have still been under observation for less than 5 years. Two patients have died from seemingly unrelated causes but no tumour was found at post-mortem, possibly because, if present, they were still too small to have been detected by crude naked-eye examination!

The κ/λ ratio of the various Bence-Jones proteins associated with the different paraprotein classes vary considerably as shown in Table 5, which is largely taken from Fine (64), supported by data of my own for IgM and IgD.

Rheumatoid factors are generally κ and amongst patients with primary cold haemagglutinin syndrome only one case of type λ Bence-Jones protein has been reported and that twice! In all other cases the Bence-Jones proteins have belonged to the κ class and indeed probably the same subclass (65), which is perhaps not unexpected since they represent antibodies against the I antigen of human erythrocytes.

Nearly all the IgD paraproteins have been λ class (three of those originally reported as κ chains are now known to have been mistyped!) and this is in keeping with the 87% incidence of λ type in normal plasma cells which contain IgD (13). It is also possible that the λ chain associated with IgD is of an unusual subclass(es) which may make typing of IgD difficult with commercial antisera.

It was formerly believed that κ and λ myelomata carried different prognoses, but the long follow-up of patients in the Medical Research Council trials has shown this not to be so. It was found, however, that κ-type Bence-Jones proteinuria was associated with more early deaths from renal failure. This is responsible for the better initial prognosis of patients with λ-type paraproteinaemias and their apparently better response to treatment during the first year. Urinary output of λ chains was nearly twice as high as for κ chains, and in the second year after diagnosis more deaths occurred in patients with λ-type paraproteinaemia. By three years there was no significant difference in the prognosis between the κ and λ types.

Amyloid

Amyloid is the result of deposition, in the tissues, of a homogeneous eosinophilic substance and often causes death by interfering with the function of vital organs such as the kidneys, heart and gut. It has become clear, with the use of modern methods for definition and recognition [reviewed by Hobbs (66)], that there are three types of precursor proteins which can become deposited or polymerized within tissues to produce amyloid. These three proteins were originally called A, B, C by Benditt. This classification was revised at a recent meeting in Helsinki where it was agreed to call the single identified A protein, AA and the two B proteins Aκ or Aλ according to the nature of their light chains. (This was an unfortunate choice as amongst paraprotein workers the terms Aκ and Aλ are often used as slang for paraproteins of the IgAκ and IgAλ types respectively.)

The C proteins of Benditt await further definition but currently they seem to be distinct from AA and Aκ and Aλ. They are frequently associated with the APUD cells derived embryologically from the neural crest (e.g. the C cells of the thyroid responsible for calcitonin production) and characteristically amyloid of this type is only deposited alongside such cells.

The big breakthrough in the study of amyloid came when Glenner *et al.* (35) first prepared amyloid extracts of high purity and representing up to 70% of the original amyloid material. Glenner *et al.* (35) proceeded to prove by sequence studies that the major protein content of the purified amyloid was almost identical with that of the known Bence-Jones proteins. It has been found, moreover, that B-amyloid is more frequently of the λ type which is in keeping with the known tendency for λ light chains to form polymers more readily than those of κ type.

The paraprotein classes associated with amyloid in a series of patients examined in my own laboratory are shown in Table 6.

Obviously not all paraproteins cause amyloid; only some 20% do. But although Bence-Jones proteinuria only has the highest association, typical B amyloid can also occur when Bence-Jones protein concentrations are below the detection limit of 10 mg/litre in urine or serum.

Table 6. *Paraprotein classes and amyloid*

This Table incorporates light-chain typing data from the literature plus 55 personal patients with amyloid and compares them with 1300 consecutive paraproteins typed in my laboratory (excluding selected referrals).

% of total						
IgG	IgA	IgM	IgD	Bence-Jones protein only	Number of patients	κ/λ ratio
30	11	11	3	45	Amyloid: 135	1.25
52	24	8	1	15	Total: 1300	2.0

Glenner and his colleagues confirmed the prediction by Osserman *et al.* (67) that Bence-Jones proteins give rise to amyloid because they have a direct affinity for the tissues in which they are deposited. B amyloid is often 'atypical' in distribution because of the weak binding in V_L portions and shows the variability in tissue involvement that might be expected from V_L affinities. The variability of B amyloid contrasts with the 'typical' distribution of amyloid due to A protein which, in sequence studies to date, have all been identical.

The relative frequency of A and B amyloid is only just beginning to be verified by modern methodology, but in a large personal series of 115 patients with amyloid, paraproteins were clearly associated with 42% and clearly not so in 52%.

Because A protein seems to be formed excessively in chronic inflammatory states, and, since paraproteins may themselves be associated with excessive infection due to suppression of normal B-cell function, it is not altogether surprising that both A and B amyloid can occur together in about 5–10% of patients nor that in some patients with paraproteins, only A amyloid is found. Conversely, chronic inflammatory states may provoke long-continued B-cell overactivity with an up to threefold increase in the incidence of paraproteins. Consequently the amyloid in chronic inflammatory disease states is of B type in about 5–10% of the cases. Taking such overlaps into consideration it seems

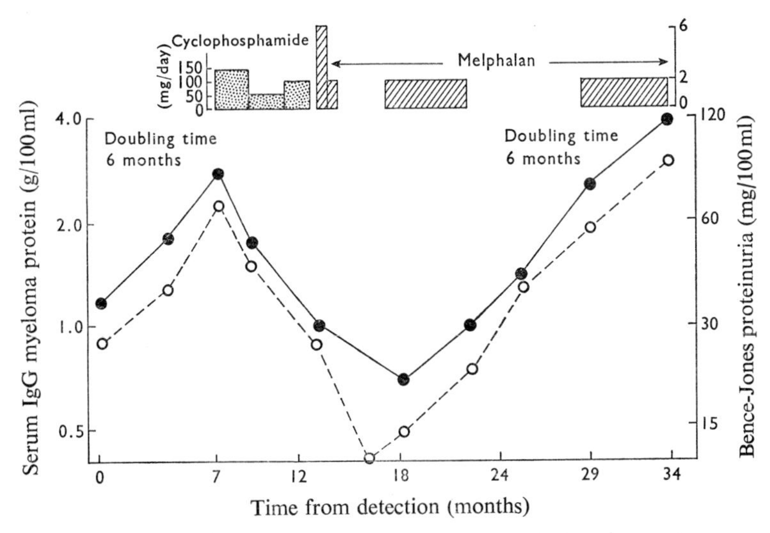

Fig. 6. '*Simple escape' from the cytotoxic treatment of myelomatosis*

The only change in the tumour with the passage of time as judged by the proteins produced has been the acquired resistance to drugs. Note that the proportion of whole IgG in the plasma (●) to Bence-Jones protein in the urine (○) and the doubling rate remained the same throughout the period of treatment.

that about half the amyloid causing clinical problems in Britain is of the B type.

Because amyloid can be the presenting feature in a patient, without its underlying cause being obvious, the term 'Primary amyloid' has become accepted. Nevertheless, of 40 patients given this apology of a diagnosis, but subsequently investigated further, no less than 35 had paraproteins and 33 eventually declared a malignant tumour of one of the types listed in Table 1. These findings bear out the general malignant nature of Bence-Jones proteins and justifies attempts in such patients to prevent further synthesis of the precursor proteins.

Plate 3(*a*) shows how Bence-Jones proteinuria was eliminated in just such a patient (68) who is currently alive and well with improved renal function. Thus some regression of deposited amyloid is possible. In general this would seem to be likely only in early cases, which emphasizes the need for adequate chemical analysis of urine from all patients in whom a diagnosis of amyloid is made or even contemplated. It should be remembered, moreover, that 8% of all cases of adult nephrosis are due to amyloid which means that it is far from rare.

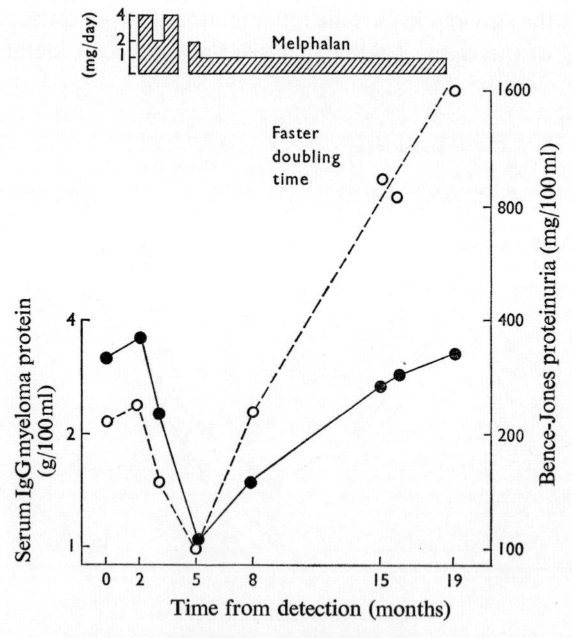

Fig. 7. '*Bence-Jones escape*'

A more than eight-fold 'excess' of Bence-Jones proteinuria, to over 24g/day, developed during the relapse and was inexplicable solely on the basis of renal tubular reabsorption and its subsequent failure. The rate of increase of Bence-Jones proteinuria (○) was no longer parallel to the parent IgG (●); i.e. a primitive subclone could have been emerging.

Mutation

Because of the ability to harvest Bence-Jones proteins from urine and the large amount of work done in this field, it has been possible, in man, actually to observe somatic mutations occurring. In the Medical Research Council myeloma trials already referred to, both serum and urine were monitored during the treatment of the patients. In about half the patients who initially had serum paraproteinaemia with Bence-Jones proteinuria and responded to treatment, the two markers of disease varied in parallel (Fig. 6) indicating that no significant change had occurred in protein production by the underlying cell clone as a result of the drug therapy. In the remaining half of the patients, however, the ratio of paraproteins in the plasma to Bence-Jones proteins in the urine clearly changed. Plate 4(*a*) illustrates, most convincingly,

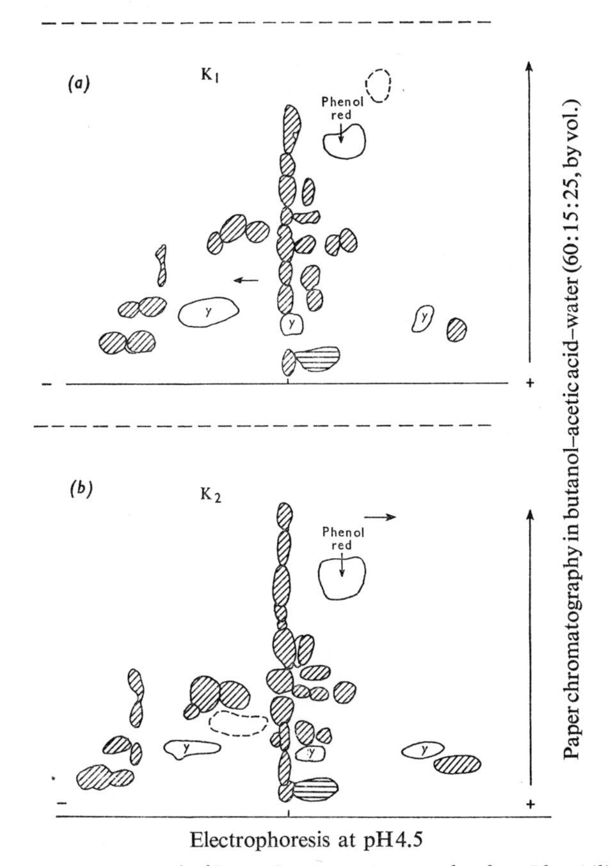

Fig. 8. '*Fingerprints' of Bence-Jones proteins* κ_1 *and* κ_2 *from Plate* 3(*b*)

The arrows and dashed lines indicate the single difference between the two peptide maps which was due to a single amino acid change (lysine for glutamine).

the appearance of excess of light chains in the urine of a patient, which 'fingerprint' comparisons showed were identical with the light chains originally fractionated from the patients paraprotein; put differently, immunoglobulin synthesis which had been balanced before treatment became unbalanced during treatment. This transformation was witnessed in three patients.

The quantitative changes shown in Fig. 7 indicate that a pre-existing imbalance can be worsened by treatment and cannot be explained solely in terms of changed renal function. I coined the term 'Bence-Jones escape' to describe these changes which occur in 35% of relapsing patients and have always so far been associated with an increased growth rate of the tumour.

In 5% of the patients a true second Bence-Jones protein emerged during treatment (Plate 4*b*) which was not due to a change in monomer/dimer ratio although this also can occur (23). 'Fingerprint' studies in three such patients produced results like those shown in Fig. 8 in which a single polypeptide change was found. Amino acid analyses of the relevant peptides showed a single amino acid change (e.g. lysine for glutamate) compatible with a change of a single nucleotide (e.g. adenine for guanine) in the DNA codon. Thus one-point mutations have been shown to occur during man's lifespan. Changes of this type have so far only been witnessed during attempts to poison the DNA of malignant cells and may be the price we have to pay for the other benefits of interfering with the thread of life.

Conclusion

It would be inappropriate to try and summarize further the vast amount of work that has been done in this area since MacIntyre first observed the urinary protein to which Bence Jones added his name. I believe that both of them would have been fascinated to see how this, the first marker of a malignant tumour, has been used both to help individual patients and to unravel some of the mysteries of clinical and fundamental biochemistry.

References

(1) Clamp, J. R. (1967) Some aspects of the first recorded case of multiple myeloma. *Lancet* **ii**, 1354–1356
(2) J. von Rustizky, (1873) Multiple myeloma. *Z. Chirurg.* **3**, 102–110
(3) Kahler, O. (1889) On the symptomatology of multiple myeloma. Observations of albumosuria. *Prager Med. Wochenschr.* **14**, 33–34 (in German)
(4) Snapper, I. & Kahn, A. (1971) *Myelomatosis*, pp. 1–9, Karger, Basel
(5) Hobbs, J. R. (1974) Laboratory screening and monitoring of cancer. *Lancet* **ii**, 1305–1307
(6) Hobbs, J. R. (1974) Monitoring myelomatosis. *Arch. Intern. Med.* **135**, 125–130
(7) Putnam, F. W. (1960) *The plasma proteins* vol 2, pp. 345–406, Academic Press, New York and London
(8) Marchalonis, J. J. & Nossal, G. J. V. (1968) Electrophoretic analysis of antibody produced by single cells. *Proc. Nat. Acad. Sci. U.S.* **61**, 860–867

(9) Wang, A. C., Wilson, F. K., Hopper, J. E., Fudenberg, H. H. & Nisonoff, A. (1970) Evidence for control of synthesis of the variable regions of the heavy chains of immunoglobulins G and M by the same gene. *Proc. Nat. Acad. Sci. U.S.* **66**, 337–343
(10) Talmage, D. W. (1969) The nature of the immunological response. In *Immunology and Development* (Adinolfi, M., ed,), pp. 1–26, Heinemann, London
(11) Bernier, G. M. & Cebra, J. J. (1965) Frequency distribution of α, γ, κ and λ polypeptide chains in human lymphoid tissues. *J. Immunol.* **95**, 246–253
(12) Takahashi, M., Yagi, Y. & Pressman, D. (1969) Preparation of fluorescent antibody reagents monospecific to light chains of human immunoglobulins. *J. Immunol.* **102**, 1268–1273
(13) Pernis, B., Governa, M. & Rowe, D. S. (1969) Light chain types in plasma cells that produce IgD. *Immunology* **16**, 685–689
(14) Hijmans, W., Schuit, H. R. E. & Radl, J. (1973) Deviations in the balance of intracellular heavy- and light-chain determinants in human plasma cells. *Prot. Biol. Fluids Proc. Colloq.* **20**, 181–184
(15) Bernier, G. M. & Putnam, F. W. (1964) Polymerism, polymorphism, and impurities in Bence-Jones proteins. *Biochim. Biophys. Acta* **86**, 295–308
(16) Hobbs, J. R. (1966) The detection of Bence-Jones proteins. *Biochem. J.* **99**, 15P
(17) Harrison, G. A. (1947) *Chemical Methods in Clinical Medicine*, 3rd. edn., Churchill, London
(18) Jirgensons, B., Ikenaka, T. & Gorguraki, V. (1959) Concerning chemistry and testing of Bence-Jones proteins. *Clin. Chim. Acta* **4**, 876–882
(19) Cohen, E. & Raducha, J. J. (1962) Detection of urinary Bence-Jones protein by means of *p*-toluenesulfonic acid (T.S.A.). *Amer. J. Clin. Pathol.* **37**, 660–663
(20) Bradshaw, T. R. (1906) The recognition of myelopathic albumin in urine. *Brit. Med. J.* **ii**, 1442–1444
(21) Kohn, J. (1973) Cellulose acetate electrophoresis. In *Laboratory Medicine* (G. J. Race, ed.), vol. 12B, pp. 1–15, Harper and Row, London
(22) Hobbs, J. R. (1965) A staining method for proteins and dextrans on cellulose acetate *Nature (London)* **207**, 292–293
(23) Virella, G., Pires, M. T. & Coelho, I. M. (1974) Analytical characterization of the urinary proteins from sixty patients with monoclonal gammapathies. *Clin. Chim. Acta* **50**, 63–75
(24) Grant, G. H. & Everall, P. H. (1965) Gel immunofiltration. A new technique for the qualitative analysis of serum proteins. *J. Clin. Pathol.* **18**, 654–659
(25) Hobbs, J. R. (1966) Disturbances of the immunoglobulins. *Sci. Basis Med.* 106–127
(26) Hobbs, J. R. & Jacobs, A. (1969) A half-molecule GK plasmacytoma. *Clin. Exp. Immunol.* **5**, 199–207
(27) Laurell, C. B. (1970) Complexes formed *in vivo* between immunoglobulin light chain kappa, prealbumin, and/or alpha$_1$-antitrypsin in myeloma sera. *Immunochemistry* **7**, 461–465
(28) Pesce, A. J., Boreisha, I. & Pollak, V. E. (1972) Rapid differentiation of glomerular and tubular proteinuria by sodium dodecyl sulphate polyacrylamide gel electrophoresis. *Clin. Chim. Acta* **40**, 27–34
(29) Waldmann, T. A., Strober, W. & Mogielnicki, R. P. (1972) The renal handling of low molecular weight proteins. *J. Clin. Invest.* **51**, 2162–2174
(30) Tan, M. & Epstein, W. V. (1965) A direct immunologic assay of human sera for Bence-Jones proteins (L-chains). *J. Lab. Clin. Med.* **66**, 344–356
(31) Putnam, F. W. (1973) Principles of immunoglobulin structure in relation to aberrations in protein metabolism. *Prot. Biol. Fluids Proc. Colloq.* **20**, 29–37
(32) Nussenzweig, V. & Benacerraf, B. (1967) Antihapten antibody specificity and L chain type. *J. Exp. Med.* **126**, 727–743
(33) Orlans, E. & Pearce, C. A. (1974) Antigenic determinants in the *N*-terminal halves of human κ-chains; their relation to the variable region subgroups. *Eur. J. Immunol.* **4**, 86–90

(34) Potter, M. & Lieberman, R. (1970) Common individual antigenic determinants in five of eight BALB/c IgA myeloma proteins that bind phosphoryl choline. *J. Exp. Med.* **132**, 737–751
(35) Glenner, G. G., Terry, W., Harada, M., Isersky, C. & Page, D. (1971) Amyloid fibril proteins: proof of homology with immunoglobulin light chains by sequence analyses. *Science* **172**, 1150–1151
(36) Hood, L. & Ein, D. (1968) Immunoglobulin Lambda chain structure: two genes, one polypeptide chain. *Nature* (*London*) **220**, 764–767
(37) Askonas, B. A. & Williamson, A. R. (1967) Balanced synthesis of light and heavy chains of immunoglobulin G. *Nature* (*London*) **216**, 264–267
(38) Buell, D. N. & Fahey, J. L. (1969) Limited Periods of Gene Expression in Immunoglobulin-Synthesizing Cells. *Science* **164**, 1524–1525
(39) Suzuki, I., Takahashi, M. & Kamet, H. (1969) Ultrastructural localization of heavy and light polypeptide chains in human long-term culture cells detected by peroxidase-conjugated antibodies. *Experientia* **25**, 1309–1311
(40) Putnam, F. W. & Miyake, A. (1958) Proteins in multiple myeloma. VIII. Biosynthesis of abnormal proteins. *J. Biol. Chem.* **231**, 671–684
(41) Williams, R. C., Jr., Brunning, R. D. & Wollheim, F. A. (1966) Light-chain disease, an abortive variant of multiple myeloma. *Ann. Intern. Med.* **65**, 471–486
(42) Suzuki, I. & Takahashi, M. (1969) Ultrastructure of human myeloma cells studied by peroxidase conjugated antibodies directed to human immunoglobulin component chains. *Experientia* **25**, 1307–1309
(43) Hobbs, J. R. (1969) Immunochemical classes of myelomatosis. *Brit. J. Haematol.* **16**, 599–606
(44) Solomon, A., Killander, J., Grey, H. M. & Kunkel, H. G. (1966) Low-molecular-weight proteins related to Bence-Jones proteins in multiple myeloma. *Science* **151**, 1237–1239
(45) Hobbs, J. R. (1971) Immunoglobulins in Clinical Chemistry. *Advan. Clin. Chem.* **14**, 219–317
(46) Hobbs, J. R. (1967) Paraproteins, benign or malignant? *Brit. Med. J.* **iii**, 699–704
(47) Nerenberg, S. T. (1969) Gamma globulin studies of biopsy material and serum in solitary plasmacytoma of the spine. *Cancer* (*Philadelphia*) **24**, 750–757
(48) Hurez, D., Preud'homme, J.-L. & Seligmann, M. (1970) Intracellular 'monoclonal' immunoglobulin in non-secretory human myeloma. *J. Immunol.* **104**, 263–264
(49) McLaughlin, H., Wetherly-Mein, G., Pitcher, C. & Hobbs, J. R. (1973) Non-immunoglobulin-bearing 'B' lymphocytes in chronic lymphatic leukaemia? *Brit. J. Haematol.* **25**, 7–14
(50) Solomon, A. & Fahey, J. L. (1964) Bence-Jones proteinemia. *Amer. J. Med.* **37**, 206–222
(51) Caggiano, V., Dominguez, C., Opfell, R. W., Kochwa, S. & Wasserman, L. A. (1969) IgG myeloma with closed tetrameric Bence-Jones proteinemia. *Amer. J. Med.* **47**, 978–985
(52) Wochner, R. D., Strober, W. & Waldmann, T. A. (1967) The role of the kidney in the catabolism of Bence–Jones proteins and immunoglobulin fragments. *J. Exp. Med.* **126**, 207–221
(53) Hobbs, J. R. (1972) Paraproteins and the kidney. *Kidney Int.* **1**, 428–429
(54) Morgan, C. & Hammack, W. J. (1966) Intravenous urography in multiple myeloma. *N. Engl. J. Med.* **275**, 77–79
(55) Gross, H., McDonald, H. & Waterhouse, K. (1968) Anuria following urography with meglumine diatrizoate (Renografin) in multiple myeloma. *Radiology* **90**, 780–781
(56) Horn, M. F., Knapp, M. S., Page, F. T. & Walker, W. H. C. (1969) Adult Fanconi syndrome and multiple myelomatosis. *J. Clin. Pathol.* **22**, 414–416
(57) Harris, R. I. & Kohn, J. A. (1974) Urinary cryo-Bence-Jones protein gelling at room temperature. *Clin. Chim. Acta* **53**, 233–237
(58) Berggard, I. & Edelman, G. M. (1963) Normal counterparts to Bence-Jones proteins: free L polypeptide chains of human γ-globulin. *Proc. Nat. Acad. Sci. U.S.* **49**, 330–337

(59) Gale, D. S. J., Versey, J. M. B. & Hobbs, J. R. (1974) Rocket immunoselection for detection of heavy-chain diseases. *Clin. Chem.* **20**, 1292–1294
(60) Cooper, A. G. & Hobbs, J. R. (1970) Immunoglobulins in chronic cold haemagglutinin disease. *Brit. J. Haematol.* **19**, 383–396
(61) Decastello, A. V. (1909) Beitrage zur Kenntnis der Bence-Jonesschen Albuminurie. *Z. Klin. Med.* **67**, 319–343
(62) Lindstrom, F. D., Williams, R. C. & Theologides, A. (1969) Urinary light-chain excretion in leukemia and lymphoma. *Clin. Exp. Immunol.* **5**, 83–90
(63) Dammaco, F. & Waldenström, J. (1968) Serum and light chain in benign monoclonal gammopathies, multiple myeloma and waldenström's macroglobulinaemia. *Clin. Exp. Immunol.* **3**, 911–921
(64) Fine, J. M. (1970) Study of the frequency of Kappa and Lambda light chains in 347 sera containing a monoclonal IgG, IgA, IgD, or Bence-Jones protein. *Eur. J. Clin. Biol. Res.* **15**, 199–202
(65) Cooper, A. H. & Steinberg, A. G. (1970) INV allotypes of cold agglutinin Kappa chains. *J. Immunol.* **104**, 1108 1110
(66) Hobbs, J. R. (1973) An ABC of amyloid. *Proc. Roy. Soc. Med.* **66**, 705–710
(67) Osserman, E. F., Takatsuki, K. & Talal, N. (1964) Multiple myeloma I. The pathogenesis of 'Amyloidosis'. *Semin. Hematol.* **1**, 3–85
(68) Jones, N. F., Hilton, P. J., Tighe, J. R. & Hobbs, J. R. (1972) Treatment of 'primary' renal amyloidosis with melphalan. *Lancet* **ii**, 616–619
(69) Bell, J. L. & Baron, D. N. (1968) Quantitative biuret determination of urine protein. *Proc. Ass. Clin. Biochem.* **5**, 63–64
(70) Fleischman, J. B., Pain, R. H. & Porter, R. R. (1962) Reduction of γ-globulins. *Arch. Biochem. Biophys. Suppl.* **1**, 174–180